FUN MENTAL MATH ACTIVITY BOOK

This Book Belongs to :_____

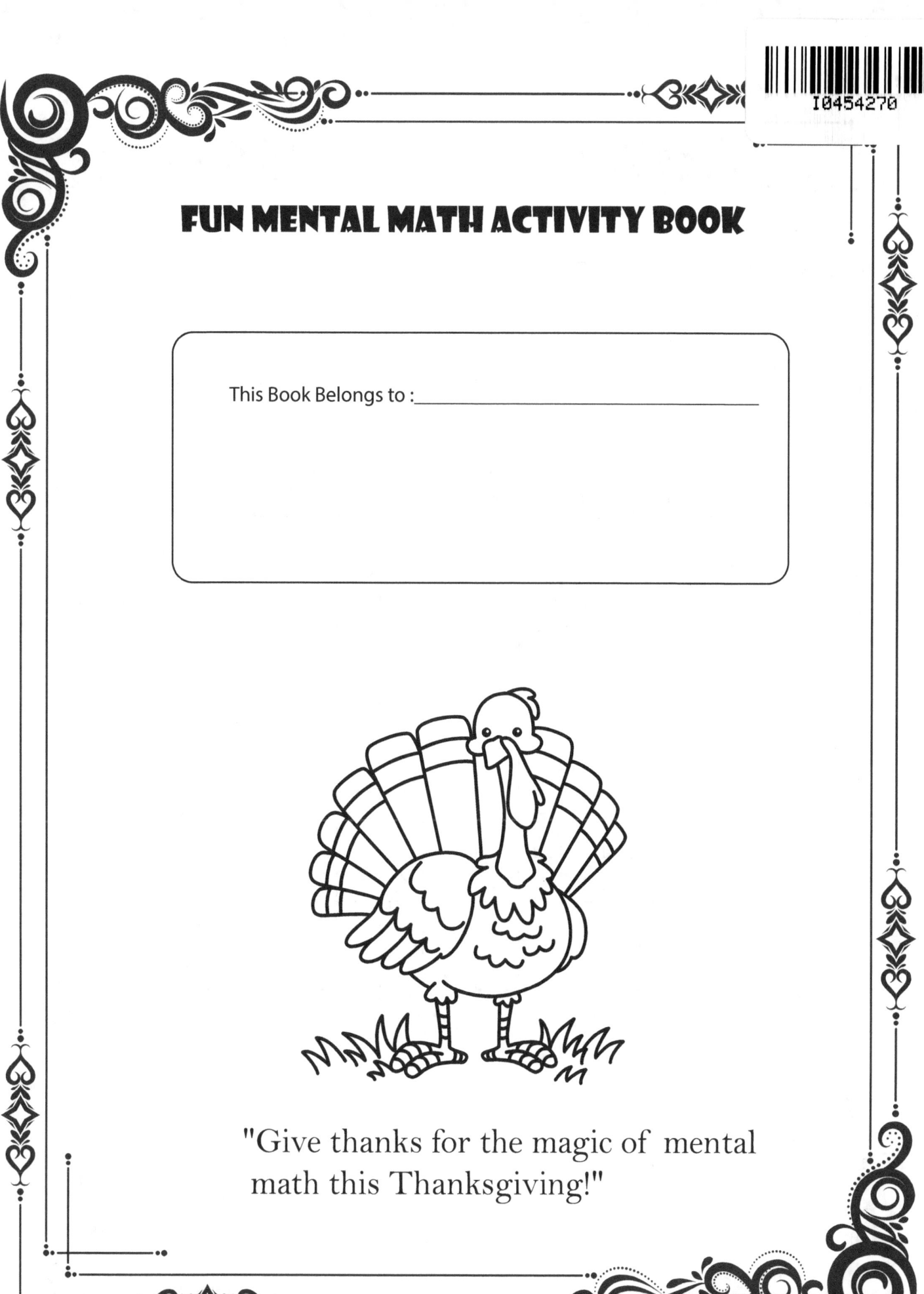

"Give thanks for the magic of mental
math this Thanksgiving!"

Across-Downs Addition

Solve.

1.

8	+	1	+	3	=	
+		+		+		+
6	+	6	+	4	=	
+		+		+		+
8	+	1	+	8	=	
=		=		=		=
	+		+		=	

2.

1	+	2	+	10	=	
+		+		+		+
5	+	10	+	2	=	
+		+		+		+
2	+	7	+	7	=	
=		=		=		=
	+		+		=	

3.

7	+	4	+	3	=	
+		+		+		+
8	+	3	+	3	=	
+		+		+		+
9	+	4	+	5	=	
=		=		=		=
	+		+		=	

4.

4	+	3	+	7	=	
+		+		+		+
6	+	3	+	6	=	
+		+		+		+
4	+	10	+	1	=	
=		=		=		=
	+		+		=	

5.

3	+	6	+	2	=	
+		+		+		+
1	+	5	+	1	=	
+		+		+		+
2	+	3	+	10	=	
=		=		=		=
	+		+		=	

6.

1	+	5	+	10	=	
+		+		+		+
1	+	7	+	2	=	
+		+		+		+
6	+	5	+	1	=	
=		=		=		=
	+		+		=	

7.

5	+	10	+	9	=	
+		+		+		+
1	+	8	+	2	=	
+		+		+		+
9	+	5	+	9	=	
=		=		=		=
	+		+		=	

8.

2	+	6	+	9	=	
+		+		+		+
4	+	2	+	1	=	
+		+		+		+
2	+	9	+	1	=	
=		=		=		=
	+		+		=	

9.

5	+	6	+	9	=	
+		+		+		+
2	+	7	+	10	=	
+		+		+		+
3	+	9	+	9	=	
=		=		=		=
	+		+		=	

10.

6	+	3	+	6	=	
+		+		+		+
5	+	10	+	9	=	
+		+		+		+
5	+	10	+	1	=	
=		=		=		=
	+		+		=	

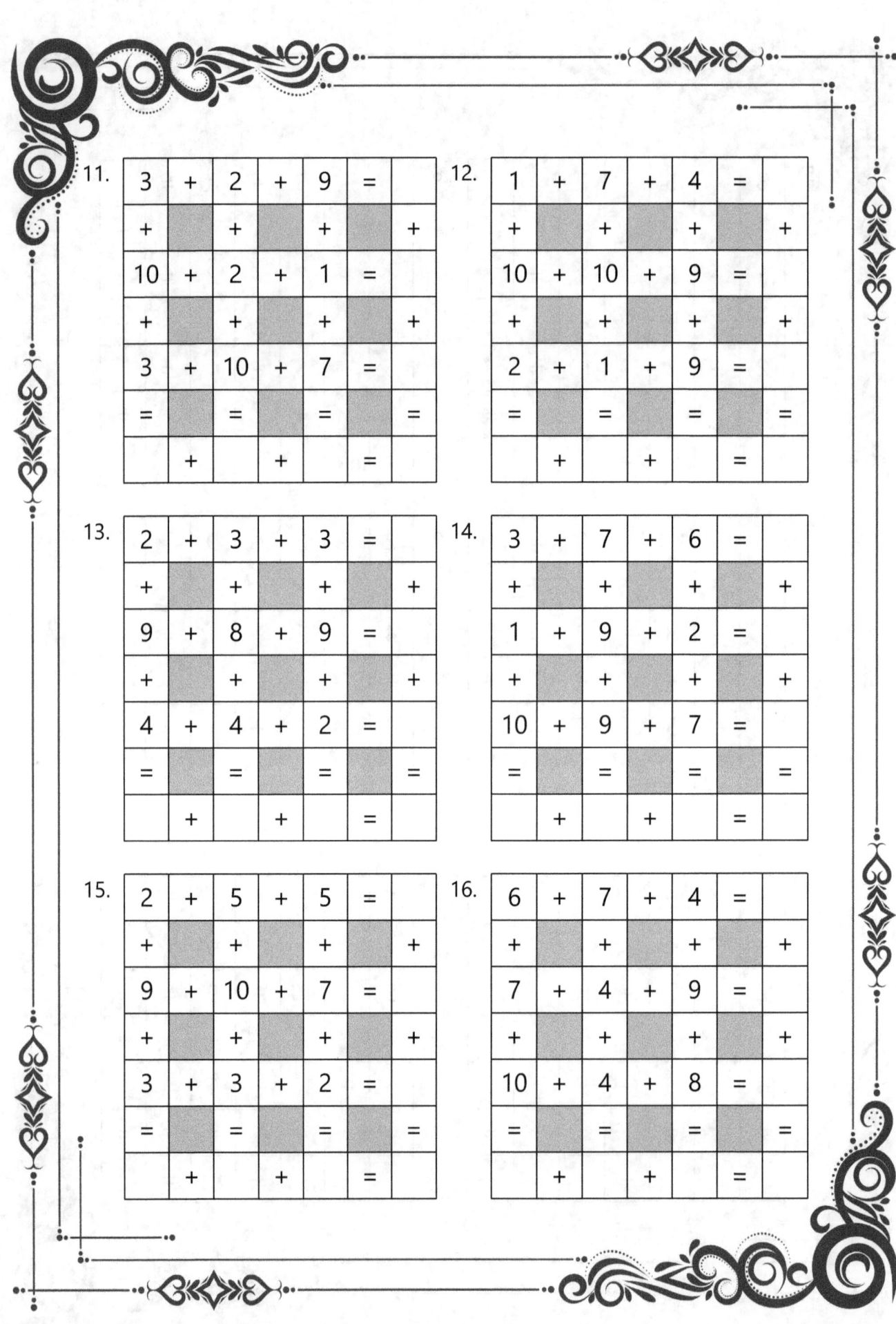

11.

3	+	2	+	9	=	
+		+		+		+
10	+	2	+	1	=	
+		+		+		+
3	+	10	+	7	=	
=		=		=		=
	+		+		=	

12.

1	+	7	+	4	=	
+		+		+		+
10	+	10	+	9	=	
+		+		+		+
2	+	1	+	9	=	
=		=		=		=
	+		+		=	

13.

2	+	3	+	3	=	
+		+		+		+
9	+	8	+	9	=	
+		+		+		+
4	+	4	+	2	=	
=		=		=		=
	+		+		=	

14.

3	+	7	+	6	=	
+		+		+		+
1	+	9	+	2	=	
+		+		+		+
10	+	9	+	7	=	
=		=		=		=
	+		+		=	

15.

2	+	5	+	5	=	
+		+		+		+
9	+	10	+	7	=	
+		+		+		+
3	+	3	+	2	=	
=		=		=		=
	+		+		=	

16.

6	+	7	+	4	=	
+		+		+		+
7	+	4	+	9	=	
+		+		+		+
10	+	4	+	8	=	
=		=		=		=
	+		+		=	

Across-Downs Subtraction

Solve.

17.

50	–	15	–	20	=	
–		–		–		–
18	–	4	–	4	=	
–		–		–		–
12	–	5	–	6	=	
=		=		=		=
	–		–		=	

18.

61	–	25	–	20	=	
–		–		–		–
18	–	8	–	7	=	
–		–		–		–
21	–	9	–	6	=	
=		=		=		=
	–		–		=	

19.

53	–	17	–	19	=	
–		–		–		–
12	–	4	–	1	=	
–		–		–		–
19	–	6	–	9	=	
=		=		=		=
	–		–		=	

20.

46	–	15	–	13	=	
–		–		–		–
18	–	10	–	1	=	
–		–		–		–
16	–	2	–	9	=	
=		=		=		=
	–		–		=	

21.

56	–	15	–	20	=	
–		–		–		–
14	–	8	–	2	=	
–		–		–		–
19	–	1	–	8	=	
=		=		=		=
		–		–		=

22.

33	–	7	–	19	=	
–		–		–		–
13	–	4	–	6	=	
–		–		–		–
15	–	2	–	10	=	
=		=		=		=
		–		–		=

23.

61	–	30	–	22	=	
–		–		–		–
23	–	10	–	10	=	
–		–		–		–
18	–	10	–	3	=	
=		=		=		=
		–		–		=

24.

50	–	18	–	13	=	
–		–		–		–
21	–	9	–	5	=	
–		–		–		–
10	–	1	–	5	=	
=		=		=		=
		–		–		=

25.

48	–	13	–	18	=	
–		–		–		–
18	–	3	–	9	=	
–		–		–		–
17	–	7	–	4	=	
=		=		=		=
		–		–		=

26.

40	–	16	–	13	=	
–		–		–		–
21	–	7	–	8	=	
–		–		–		–
9	–	2	–	3	=	
=		=		=		=
		–		–		=

27.

64	−	18	−	23	=	
−		−		−		−
23	−	7	−	9	=	
−		−		−		−
27	−	8	−	10	=	
=		=		=		=
	−		−		=	

28.

64	−	20	−	24	=	
−		−		−		−
25	−	7	−	9	=	
−		−		−		−
21	−	6	−	6	=	
=		=		=		=
	−		−		=	

29.

60	−	25	−	18	=	
−		−		−		−
16	−	7	−	7	=	
−		−		−		−
23	−	10	−	6	=	
=		=		=		=
	−		−		=	

30.

46	−	13	−	20	=	
−		−		−		−
8	−	3	−	3	=	
−		−		−		−
15	−	1	−	7	=	
=		=		=		=
	−		−		=	

31.

40	−	13	−	16	=	
−		−		−		−
21	−	9	−	6	=	
−		−		−		−
6	−	2	−	2	=	
=		=		=		=
	−		−		=	

32.

51	−	16	−	19	=	
−		−		−		−
22	−	6	−	7	=	
−		−		−		−
17	−	9	−	5	=	
=		=		=		=
	−		−		=	

Across-Downs Mixte

Solve.

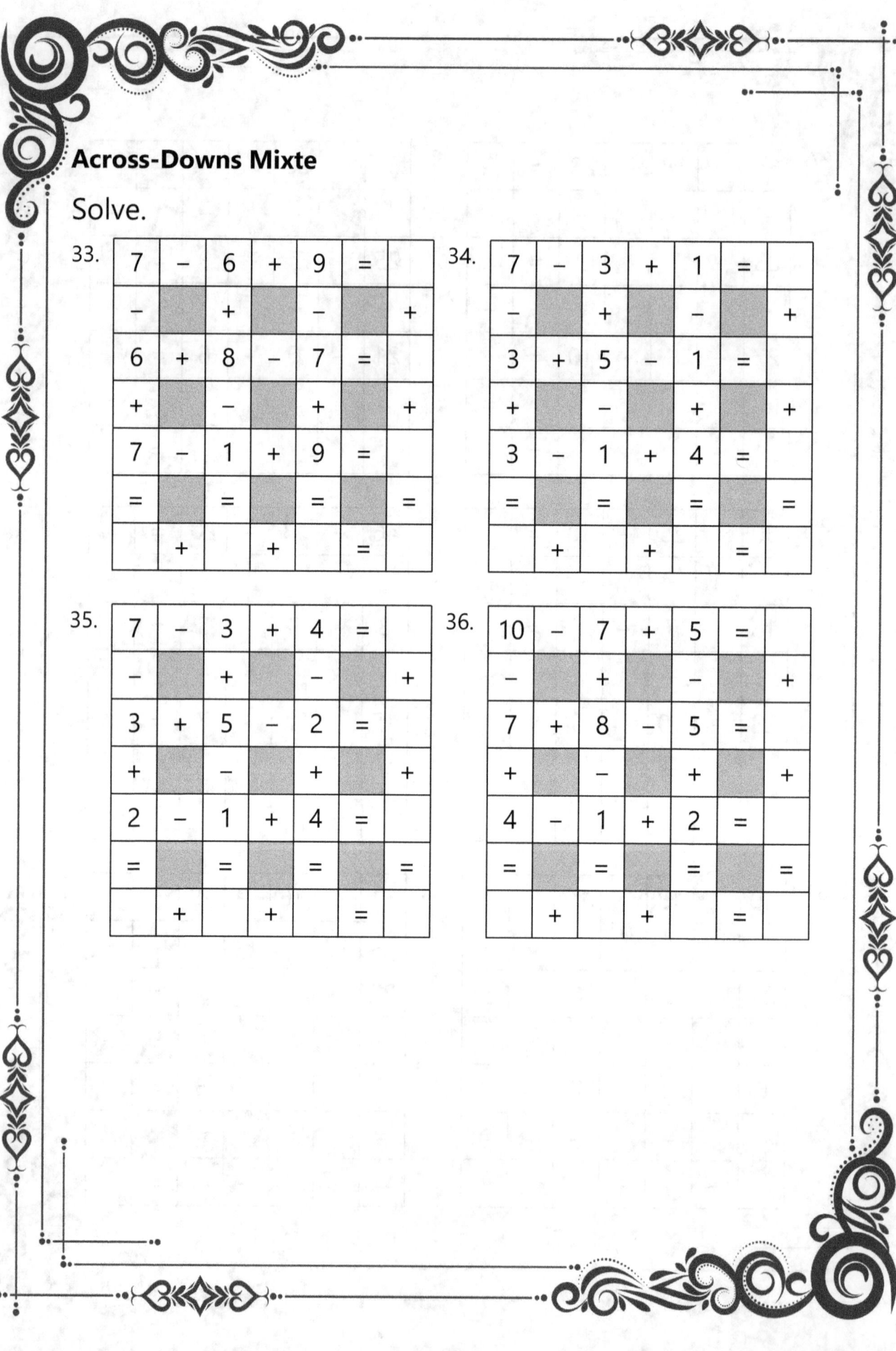

33.

7	−	6	+	9	=	
−		+		−		+
6	+	8	−	7	=	
+		−		+		+
7	−	1	+	9	=	
=		=		=		=
	+		+		=	

34.

7	−	3	+	1	=	
−		+		−		+
3	+	5	−	1	=	
+		−		+		+
3	−	1	+	4	=	
=		=		=		=
	+		+		=	

35.

7	−	3	+	4	=	
−		+		−		+
3	+	5	−	2	=	
+		−		+		+
2	−	1	+	4	=	
=		=		=		=
	+		+		=	

36.

10	−	7	+	5	=	
−		+		−		+
7	+	8	−	5	=	
+		−		+		+
4	−	1	+	2	=	
=		=		=		=
	+		+		=	

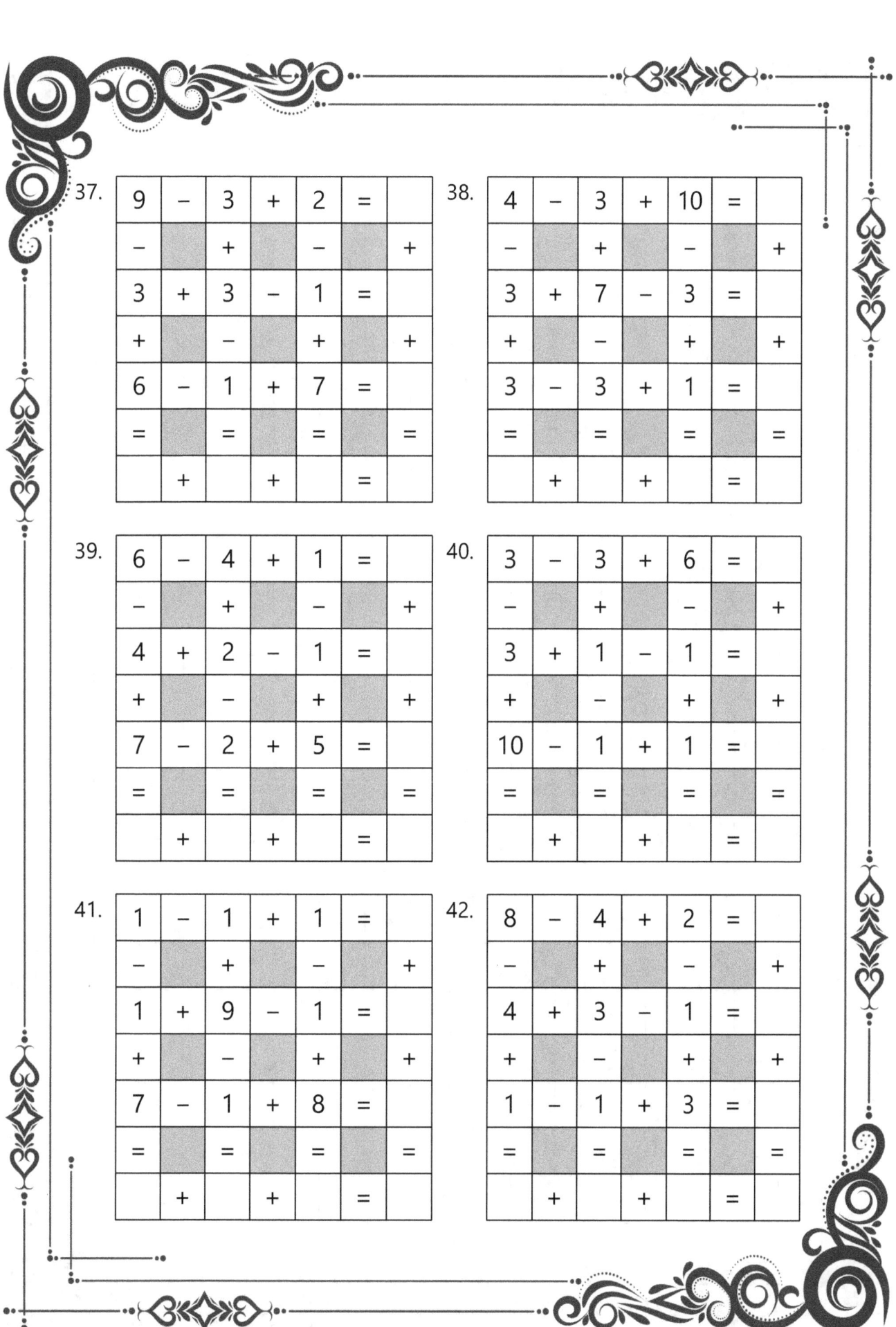

37.

9	–	3	+	2	=	
–		+		–		+
3	+	3	–	1	=	
+		–		+		+
6	–	1	+	7	=	
=		=		=		=
	+		+		=	

38.

4	–	3	+	10	=	
–		+		–		+
3	+	7	–	3	=	
+		–		+		+
3	–	3	+	1	=	
=		=		=		=
	+		+		=	

39.

6	–	4	+	1	=	
–		+		–		+
4	+	2	–	1	=	
+		–		+		+
7	–	2	+	5	=	
=		=		=		=
	+		+		=	

40.

3	–	3	+	6	=	
–		+		–		+
3	+	1	–	1	=	
+		–		+		+
10	–	1	+	1	=	
=		=		=		=
	+		+		=	

41.

1	–	1	+	1	=	
–		+		–		+
1	+	9	–	1	=	
+		–		+		+
7	–	1	+	8	=	
=		=		=		=
	+		+		=	

42.

8	–	4	+	2	=	
–		+		–		+
4	+	3	–	1	=	
+		–		+		+
1	–	1	+	3	=	
=		=		=		=
	+		+		=	

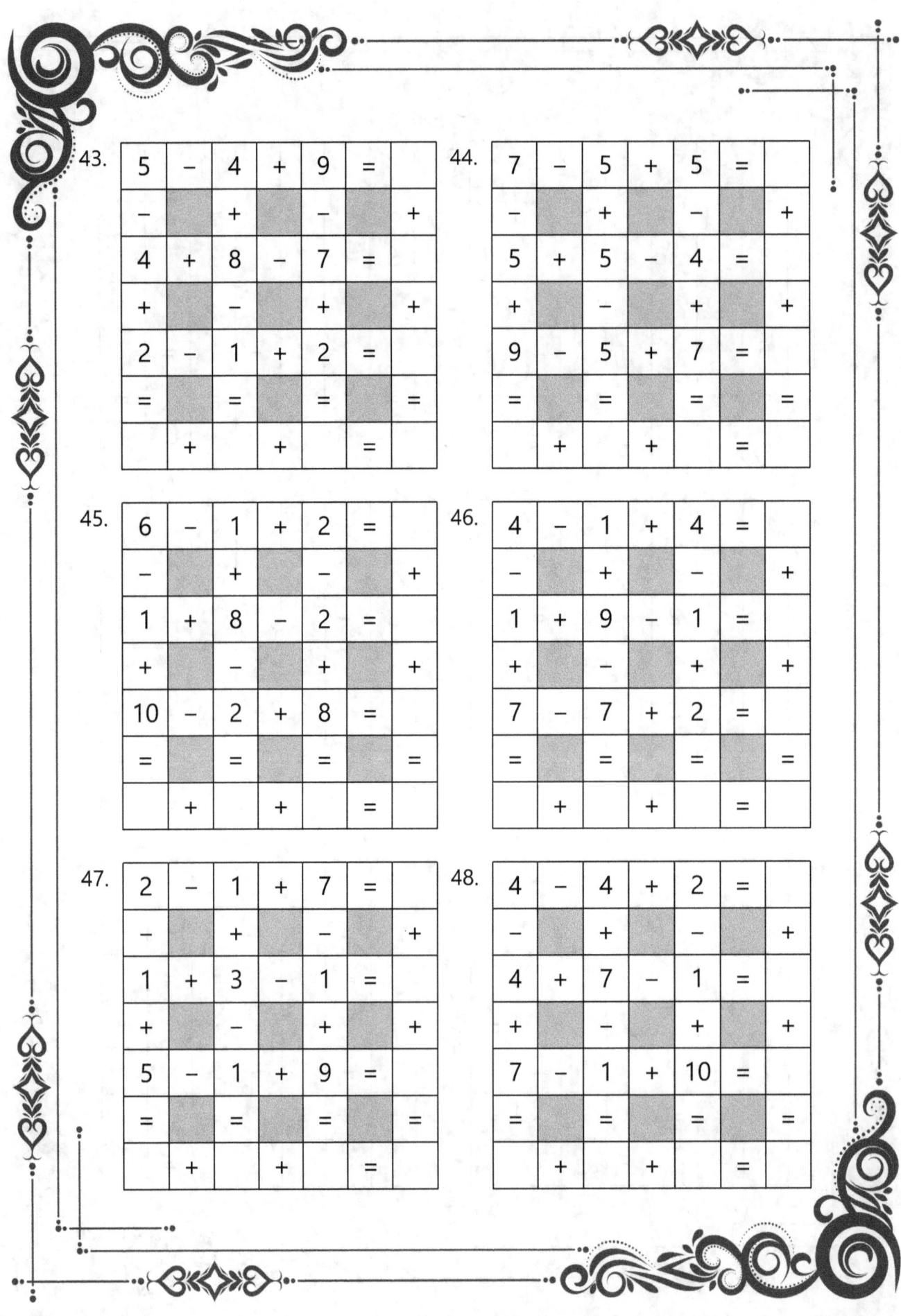

43.

5	−	4	+	9	=	
−		+		−		+
4	+	8	−	7	=	
+		−		+		+
2	−	1	+	2	=	
=		=		=		=
	+		+		=	

44.

7	−	5	+	5	=	
−		+		−		+
5	+	5	−	4	=	
+		−		+		+
9	−	5	+	7	=	
=		=		=		=
	+		+		=	

45.

6	−	1	+	2	=	
−		+		−		+
1	+	8	−	2	=	
+		−		+		+
10	−	2	+	8	=	
=		=		=		=
	+		+		=	

46.

4	−	1	+	4	=	
−		+		−		+
1	+	9	−	1	=	
+		−		+		+
7	−	7	+	2	=	
=		=		=		=
	+		+		=	

47.

2	−	1	+	7	=	
−		+		−		+
1	+	3	−	1	=	
+		−		+		+
5	−	1	+	9	=	
=		=		=		=
	+		+		=	

48.

4	−	4	+	2	=	
−		+		−		+
4	+	7	−	1	=	
+		−		+		+
7	−	1	+	10	=	
=		=		=		=
	+		+		=	

49.

5	–	3	+	8	=	
–		+		–		+
3	+	4	–	4	=	
+		–		+		+
10	–	3	+	5	=	
=		=		=		=
	+		+		=	

50.

7	–	2	+	4	=	
–		+		–		+
2	+	8	–	4	=	
+		–		+		+
6	–	1	+	1	=	
=		=		=		=
	+		+		=	

51.

7	–	3	+	3	=	
–		+		–		+
3	+	5	–	1	=	
+		–		+		+
7	–	3	+	2	=	
=		=		=		=
	+		+		=	

52.

8	–	3	+	4	=	
–		+		–		+
3	+	7	–	4	=	
+		–		+		+
5	–	3	+	8	=	
=		=		=		=
	+		+		=	

53.

1	–	1	+	4	=	
–		+		–		+
1	+	6	–	4	=	
+		–		+		+
7	–	2	+	3	=	
=		=		=		=
	+		+		=	

54.

8	–	2	+	7	=	
–		+		–		+
2	+	9	–	1	=	
+		–		+		+
6	–	1	+	8	=	
=		=		=		=
	+		+		=	

55.

10	–	2	+	1	=	
–		+		–		+
2	+	9	–	1	=	
+		–		+		+
7	–	5	+	8	=	
=		=		=		=
	+		+		=	

56.

10	–	7	+	6	=	
–		+		–		+
7	+	7	–	4	=	
+		–		+		+
10	–	4	+	2	=	
=		=		=		=
	+		+		=	

57.

1	–	1	+	8	=	
–		+		–		+
1	+	1	–	1	=	
+		–		+		+
7	–	1	+	2	=	
=		=		=		=
	+		+		=	

58.

3	–	2	+	8	=	
–		+		–		+
2	+	8	–	4	=	
+		–		+		+
7	–	3	+	4	=	
=		=		=		=
	+		+		=	

59.

8	–	6	+	6	=	
–		+		–		+
6	+	4	–	1	=	
+		–		+		+
6	–	3	+	9	=	
=		=		=		=
	+		+		=	

60.

10	–	1	+	8	=	
–		+		–		+
1	+	4	–	4	=	
+		–		+		+
7	–	3	+	3	=	
=		=		=		=
	+		+		=	

Magic Squares

Find the magic number.

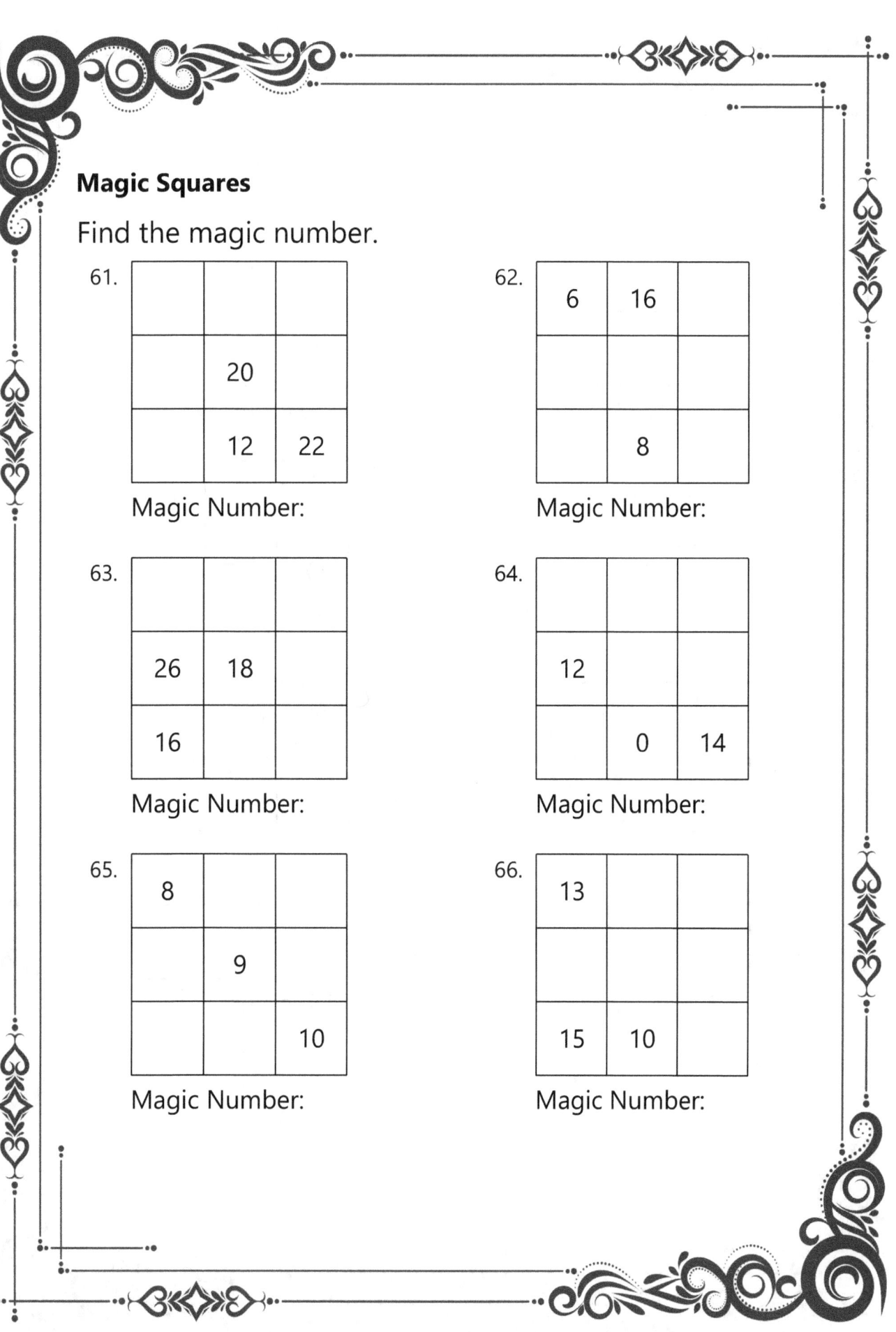

61.

	20	
	12	22

Magic Number:

62.

6	16	
	8	

Magic Number:

63.

26	18	
16		

Magic Number:

64.

12		
	0	14

Magic Number:

65.

8		
	9	
		10

Magic Number:

66.

13		
15	10	

Magic Number:

67.

	11	
17		
		14

Magic Number:

68.

	34	
	26	
28		

Magic Number:

69.

	12	
	8	
14		

Magic Number:

70.

		12
9	13	

Magic Number:

71.

		7
	6	2

Magic Number:

72.

6		
		9
	3	

Magic Number:

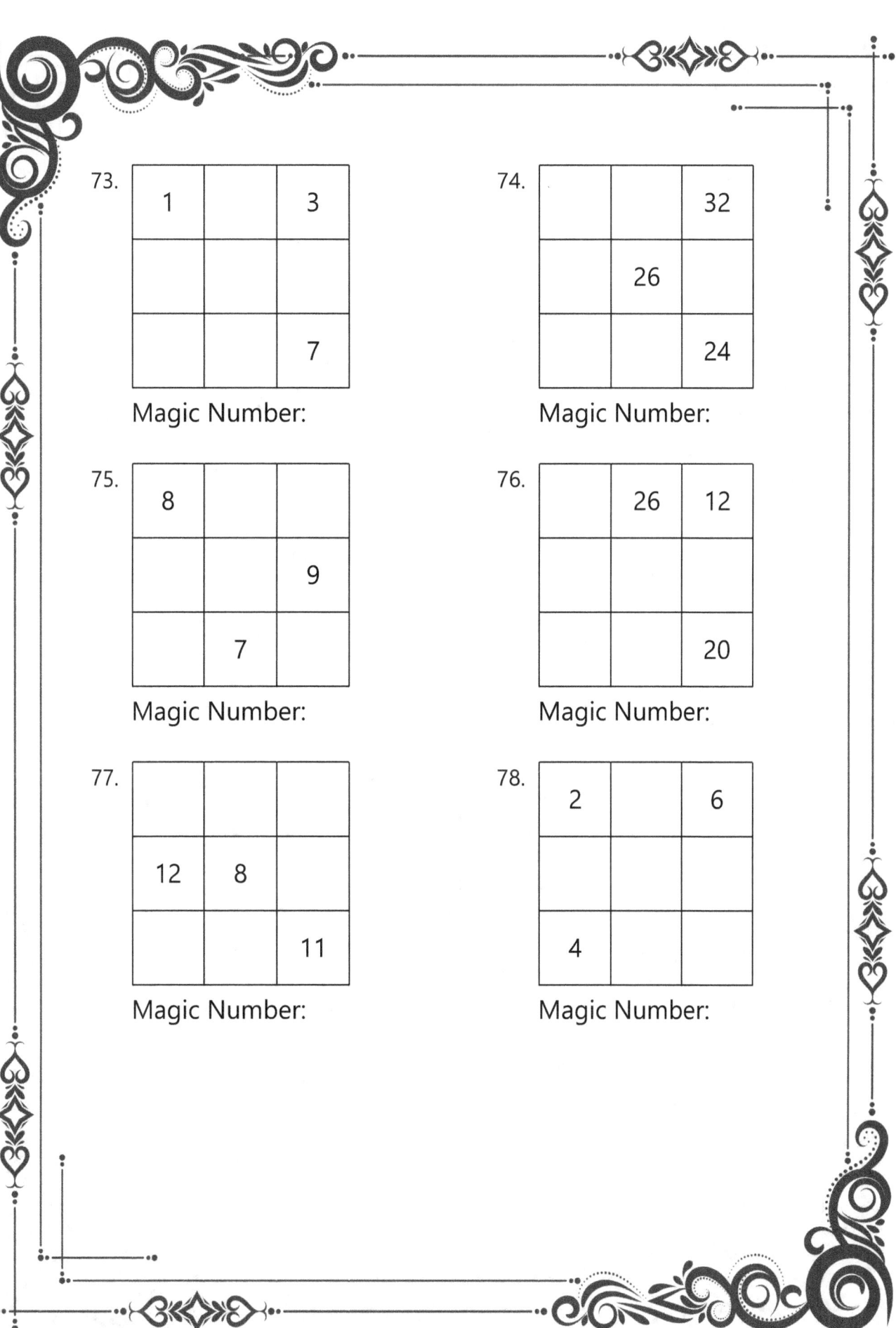

73.

1		3
		7

Magic Number:

74.

		32
	26	
		24

Magic Number:

75.

8		
		9
	7	

Magic Number:

76.

	26	12
		20

Magic Number:

77.

12	8	
		11

Magic Number:

78.

2		6
4		

Magic Number:

79.

		24
	22	
		16

Magic Number:

80.

12	7	
8		

Magic Number:

81.

	18	
	26	16

Magic Number:

82.

		16
		6
4		

Magic Number:

83.

		6
5	12	

Magic Number:

84.

		12
	8	22

Magic Number:

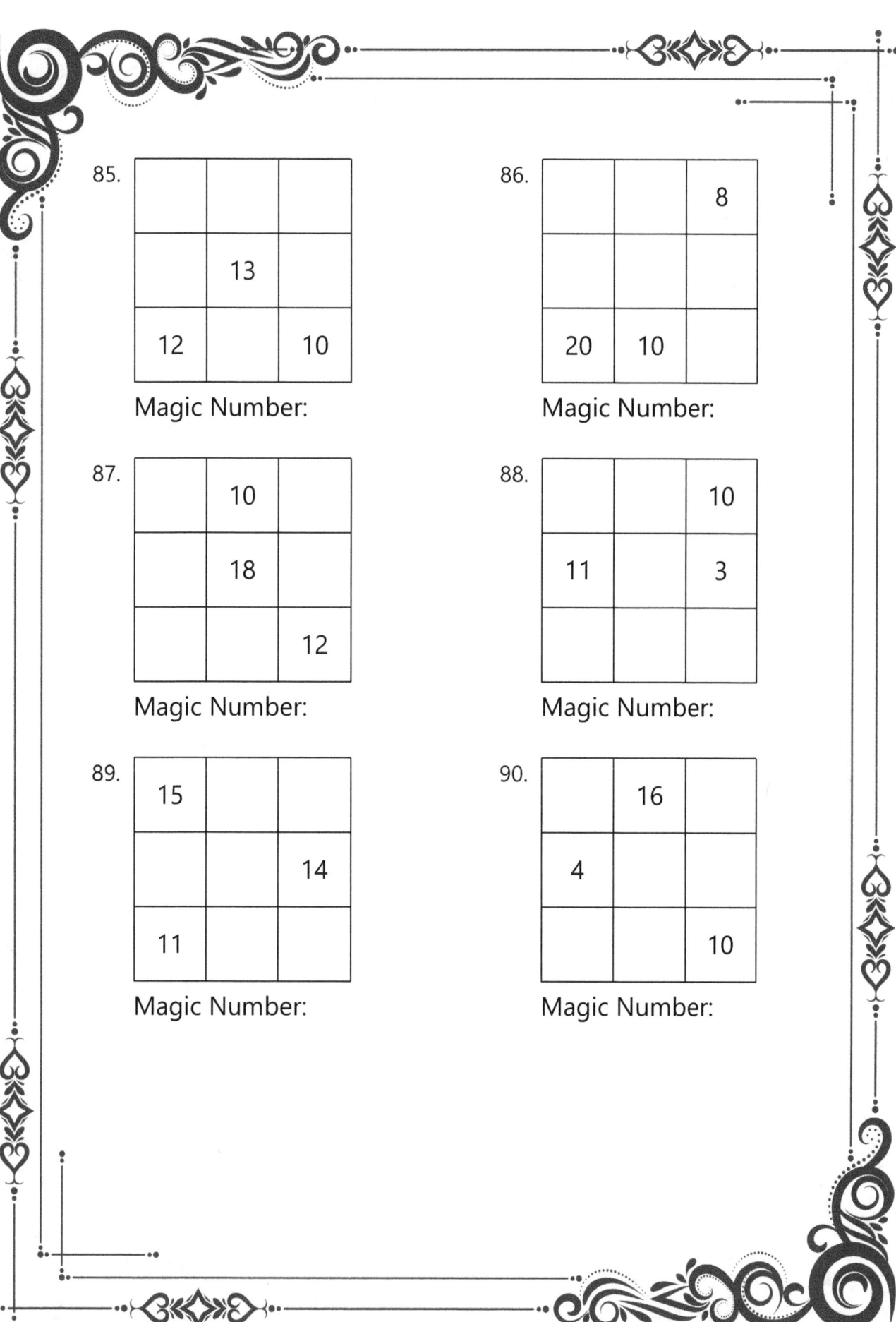

85.

	13	
12		10

Magic Number:

86.

		8
20	10	

Magic Number:

87.

	10	
	18	
		12

Magic Number:

88.

		10
11		3

Magic Number:

89.

15		
		14
11		

Magic Number:

90.

	16	
4		
		10

Magic Number:

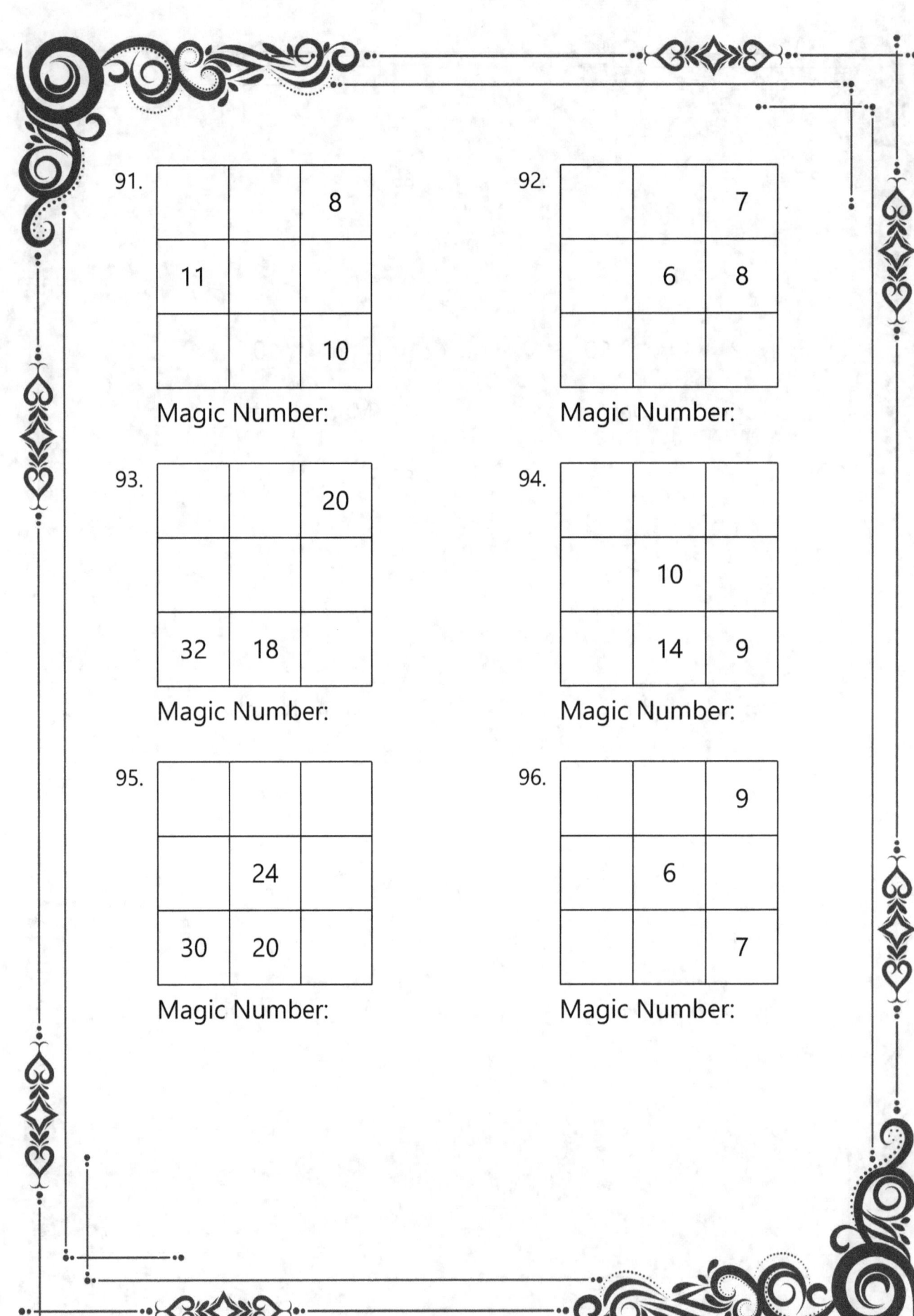

91.

		8
11		
		10

Magic Number:

92.

		7
	6	8

Magic Number:

93.

		20
32	18	

Magic Number:

94.

	10	
	14	9

Magic Number:

95.

	24	
30	20	

Magic Number:

96.

		9
	6	
		7

Magic Number:

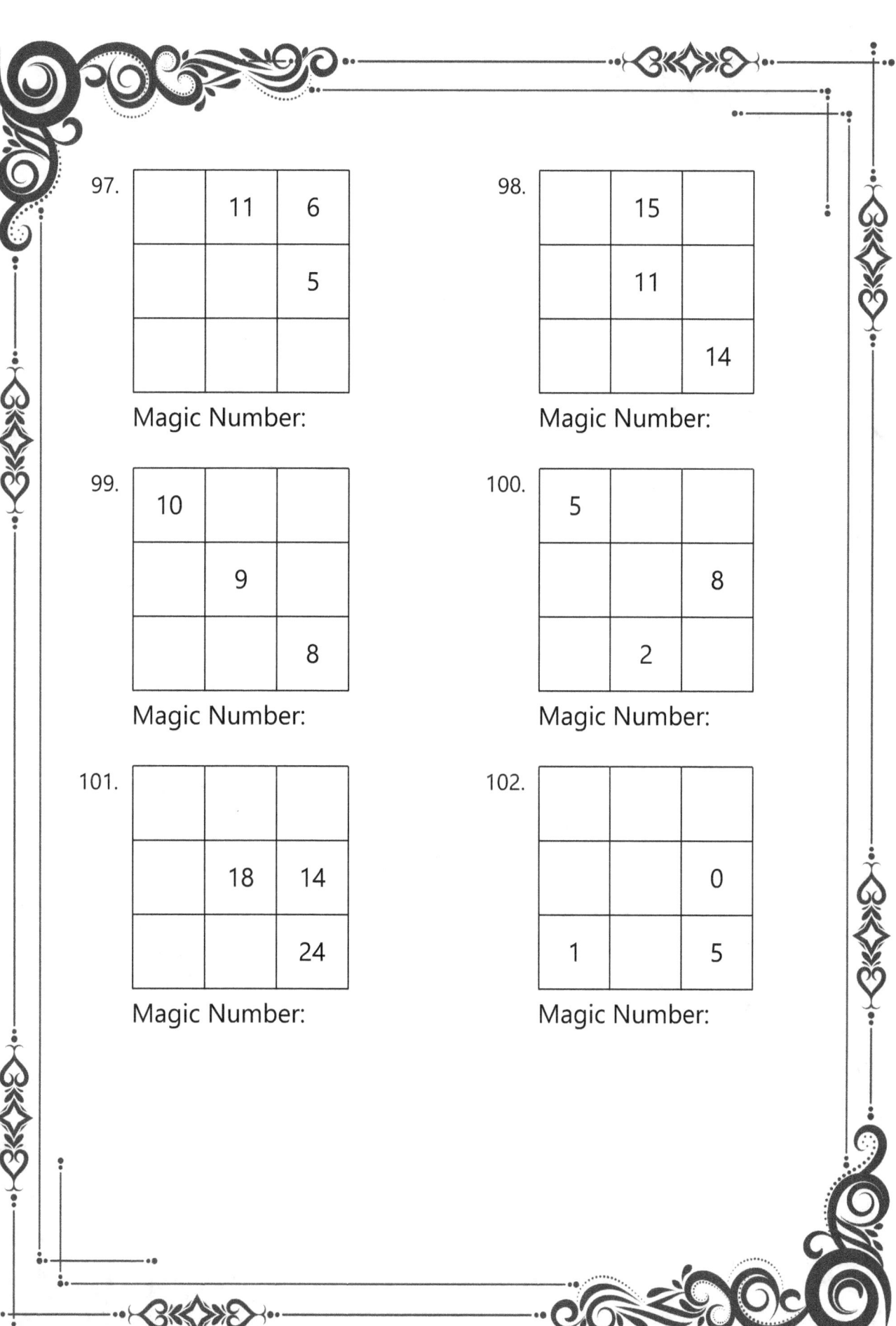

97.

	11	6
		5

Magic Number:

98.

	15	
	11	
		14

Magic Number:

99.

10		
	9	
		8

Magic Number:

100.

5		
		8
	2	

Magic Number:

101.

	18	14
		24

Magic Number:

102.

		0
1		5

Magic Number:

103.

	12	
6	20	

Magic Number:

104.

10		12
11		

Magic Number:

105.

	20	
24	16	

Magic Number:

106.

22		30
		26

Magic Number:

107.

	9	
	13	8

Magic Number:

108.

12		10
16		

Magic Number:

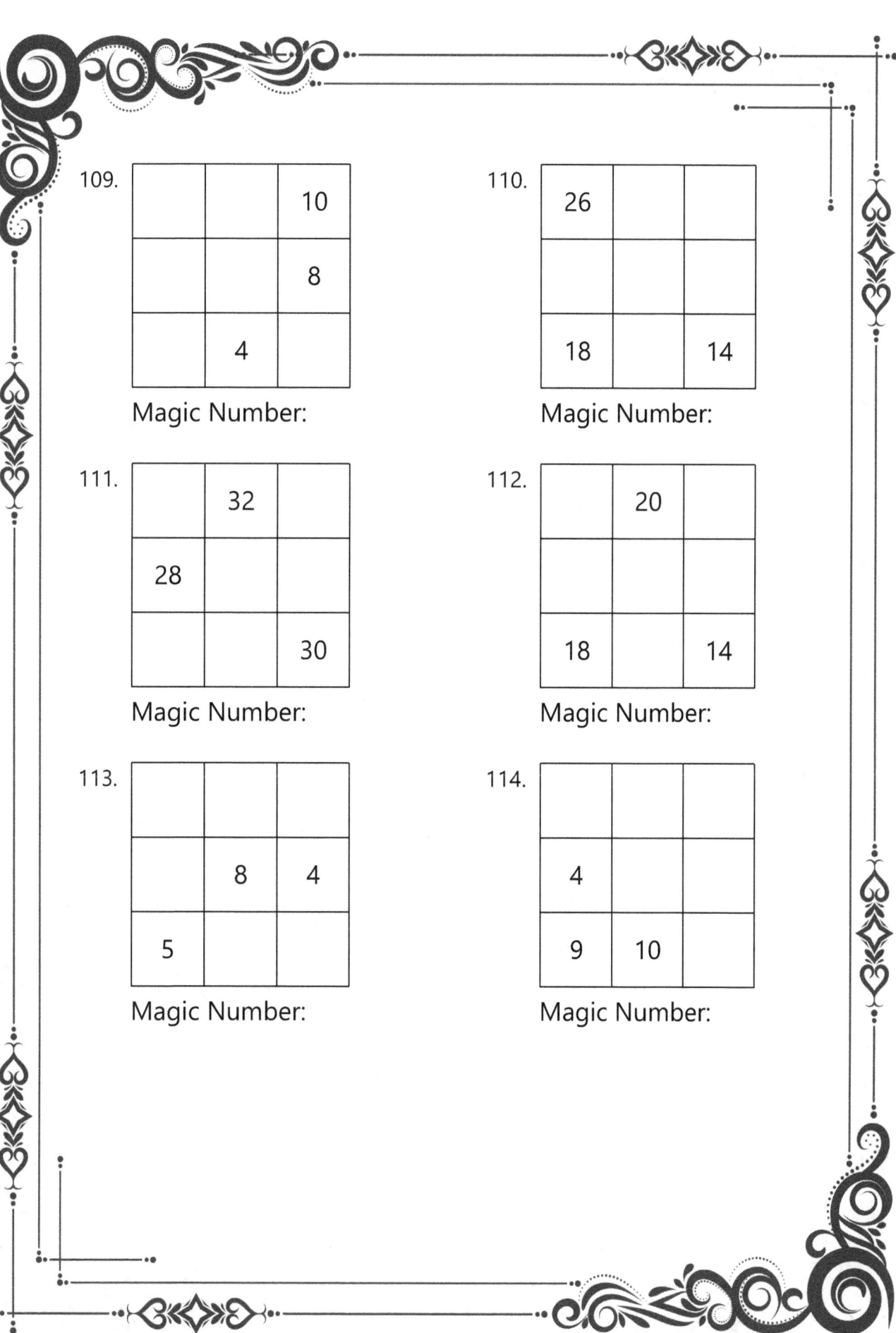

109.

		10
		8
	4	

Magic Number:

110.

26		
18		14

Magic Number:

111.

	32	
28		
		30

Magic Number:

112.

	20	
18		14

Magic Number:

113.

	8	4
5		

Magic Number:

114.

4		
9	10	

Magic Number:

115.

	8	
	12	
		14

Magic Number:

116.

	13	
		15
		10

Magic Number:

117.

	16	6
		20

Magic Number:

118.

11		
16		8

Magic Number:

119.

		14
	18	28

Magic Number:

120.

	11	
10	15	

Magic Number:

Multiplication Box

Solve.

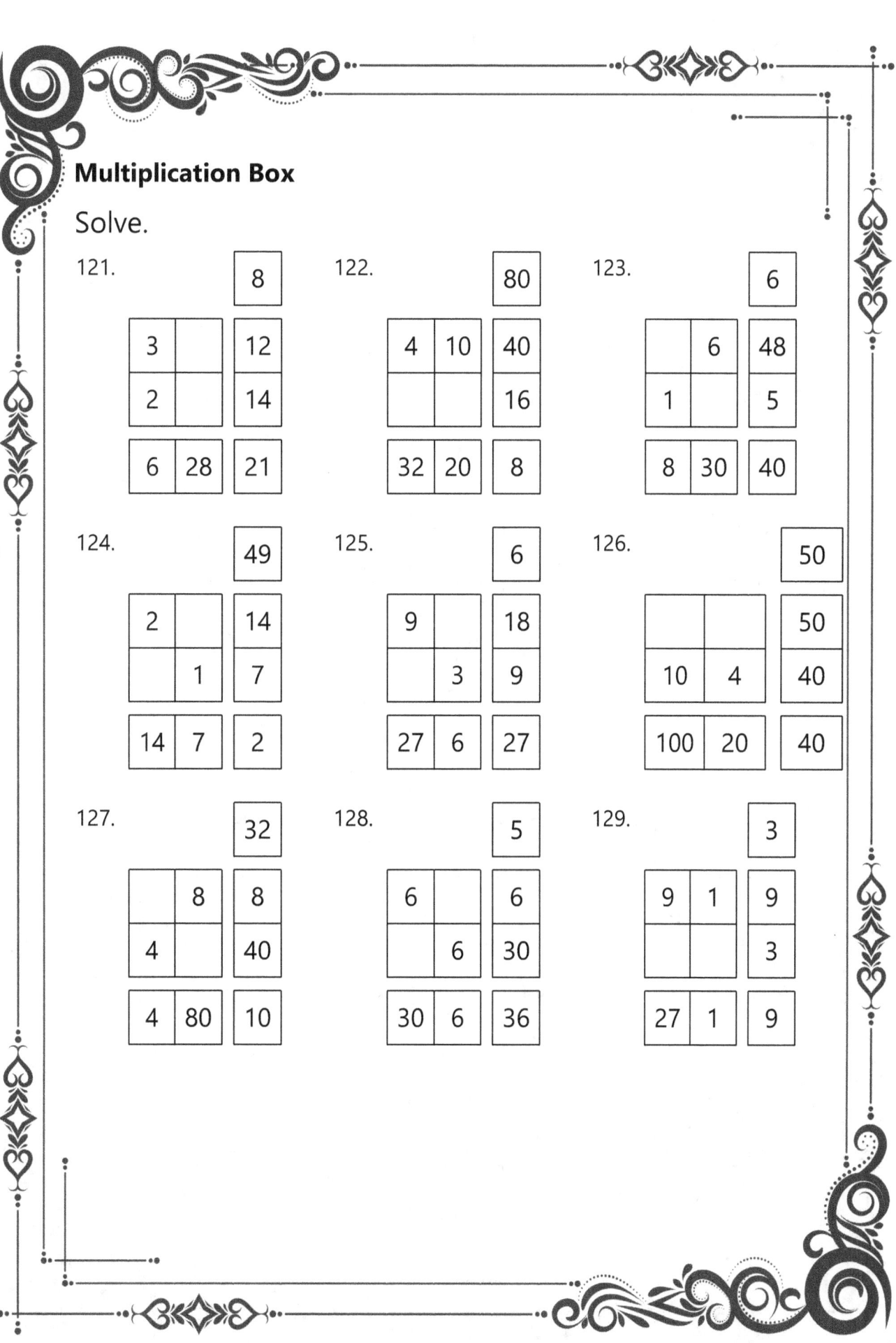

121.

		8
3		12
2		14
6	28	21

122.

		80
4	10	40
		16
32	20	8

123.

		6
	6	48
1		5
8	30	40

124.

		49
2		14
	1	7
14	7	2

125.

		6
9		18
	3	9
27	6	27

126.

		50
		50
10	4	40
100	20	40

127.

		32
	8	8
4		40
4	80	10

128.

		5
6		6
	6	30
30	6	36

129.

		3
9	1	9
		3
27	1	9

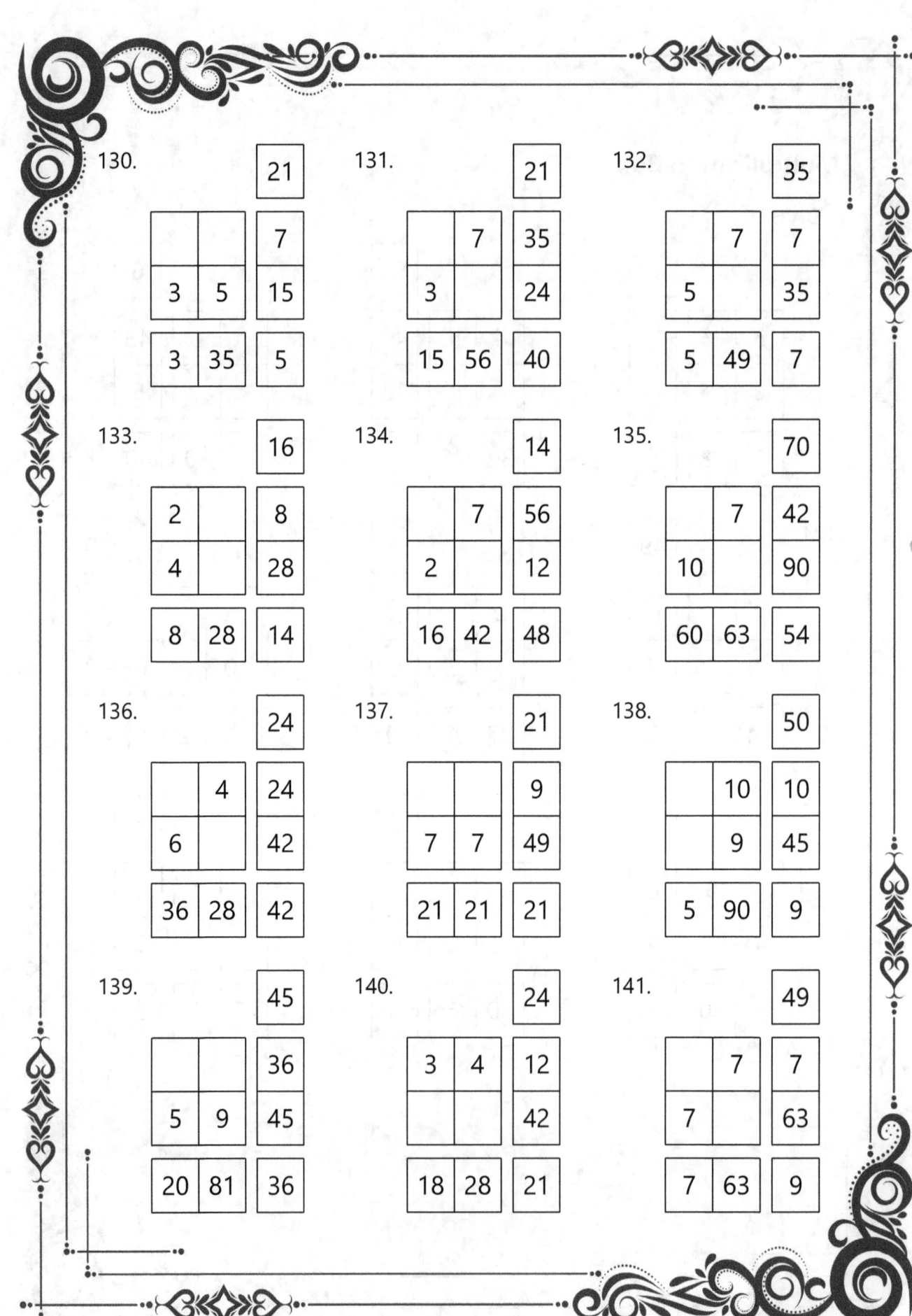

130.

		21
		7
3	5	15
3	35	5

131.

		21
	7	35
3		24
15	56	40

132.

		35
	7	7
5		35
5	49	7

133.

		16
2		8
4		28
8	28	14

134.

		14
	7	56
2		12
16	42	48

135.

		70
	7	42
10		90
60	63	54

136.

		24
	4	24
6		42
36	28	42

137.

		21
		9
7	7	49
21	21	21

138.

		50
	10	10
	9	45
5	90	9

139.

		45
		36
5	9	45
20	81	36

140.

		24
3	4	12
		42
18	28	21

141.

		49
	7	7
7		63
7	63	9

142.

		[20]
7	2	14
		80
70	16	56

143.

		[8]
2	4	8
		4
4	8	4

144.

		[32]
1		4
8		8
8	4	1

145.

		[27]
10	9	90
		9
30	27	30

146.

		[6]
1		2
	9	27
3	18	9

147.

		[24]
6		48
	7	21
18	56	42

148.

		[28]
	7	56
	2	8
32	14	16

149.

		[40]
9		45
8		16
72	10	18

150.

		[28]
9		63
	7	28
36	49	63

151.

		[10]
	5	5
	7	14
2	35	7

152.

		[72]
	9	27
8		16
24	18	6

153.

		[7]
	7	49
1		5
7	35	35

Secret Trails Addition

Find the secret trail.

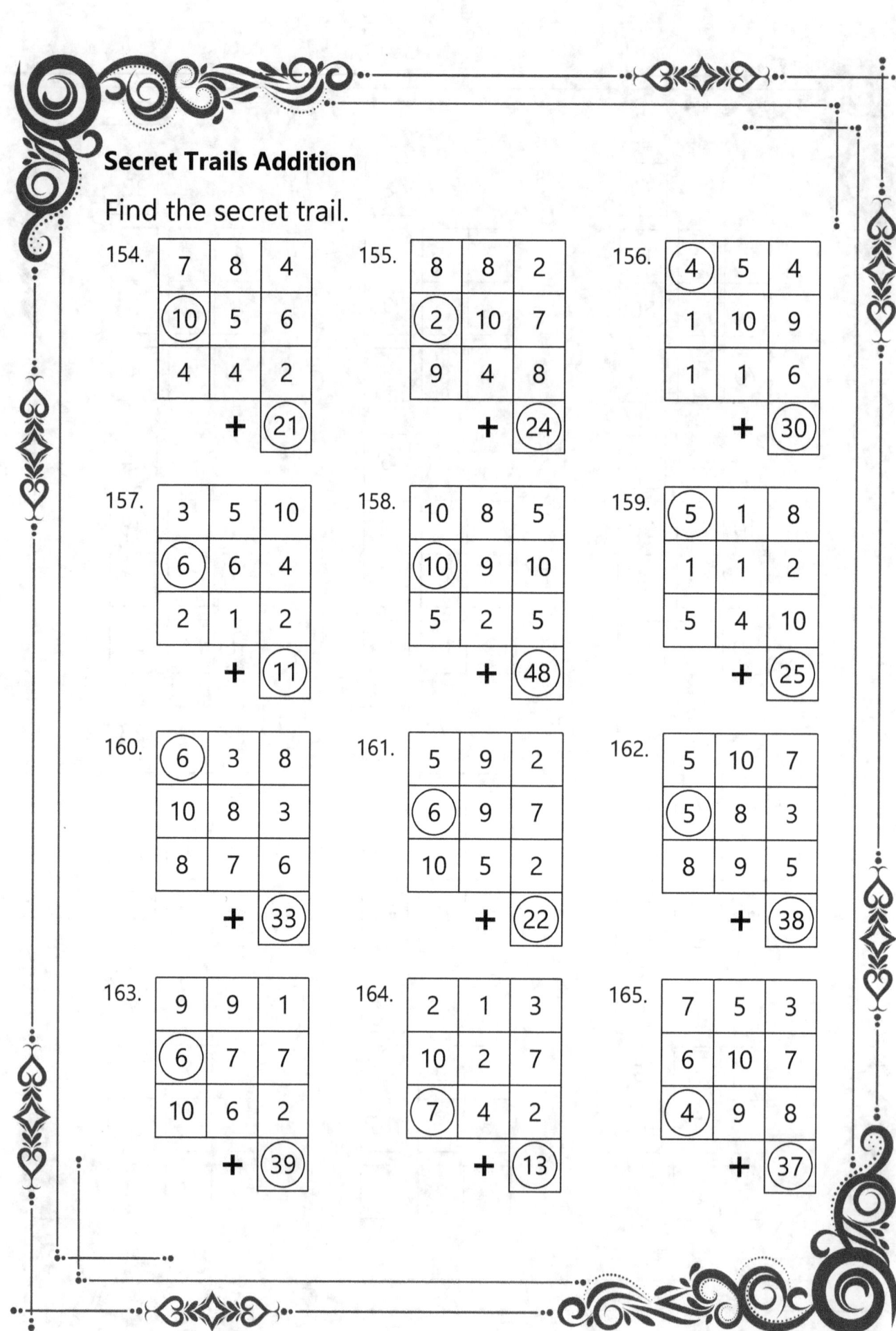

154.

7	8	4
(10)	5	6
4	4	2

+ (21)

155.

8	8	2
(2)	10	7
9	4	8

+ (24)

156.

(4)	5	4
1	10	9
1	1	6

+ (30)

157.

3	5	10
(6)	6	4
2	1	2

+ (11)

158.

10	8	5
(10)	9	10
5	2	5

+ (48)

159.

(5)	1	8
1	1	2
5	4	10

+ (25)

160.

(6)	3	8
10	8	3
8	7	6

+ (33)

161.

5	9	2
(6)	9	7
10	5	2

+ (22)

162.

5	10	7
(5)	8	3
8	9	5

+ (38)

163.

9	9	1
(6)	7	7
10	6	2

+ (39)

164.

2	1	3
10	2	7
(7)	4	2

+ (13)

165.

7	5	3
6	10	7
(4)	9	8

+ (37)

166.

4	5	8
1	4	9
⑦	2	4

+ ㉖

167.

9	4	5
10	8	8
⑧	7	1

+ ⑯

168.

10	10	7
⑦	3	1
1	8	3

+ ㉓

169.

⑧	2	3
4	7	4
2	10	7

+ ㉛

170.

9	5	10
6	10	10
⑧	8	3

+ ㊾

171.

1	2	8
③	1	7
6	5	7

+ ㉙

172.

10	9	1
①	5	2
7	2	9

+ ㊱

173.

5	1	5
⑦	2	2
2	1	4

+ ⑳

174.

3	7	3
②	4	9
5	5	10

+ ㉞

175.

④	8	1
1	6	9
1	6	8

+ ㉟

176.

1	1	6
①	9	5
7	3	6

+ ⑲

177.

8	9	7
⑥	8	8
10	7	4

+ ㉗

178.

6	4	5
③	5	2
7	2	4

+ (14)

179.

①	8	2
3	1	9
6	7	5

+ (17)

180.

7	9	1
⑦	7	2
4	8	6

+ (32)

181.

4	5	7
②	1	7
9	1	6

+ (10)

182.

8	7	5
5	3	8
⑨	9	10

+ (50)

183.

3	7	8
9	4	9
②	2	1

+ (5)

184.

10	8	8
2	6	4
⑦	8	9

+ (28)

185.

5	9	1
⑨	8	6
8	10	10

+ (51)

186.

9	10	7
①	7	2
2	5	7

+ (15)

187.

5	9	3
6	8	7
⑧	3	4

+ (47)

188.

4	2	8
2	3	2
⑤	5	6

+ (18)

189.

⑨	8	10
6	3	6
3	2	1

+ (43)

Secret Trails Subtraction

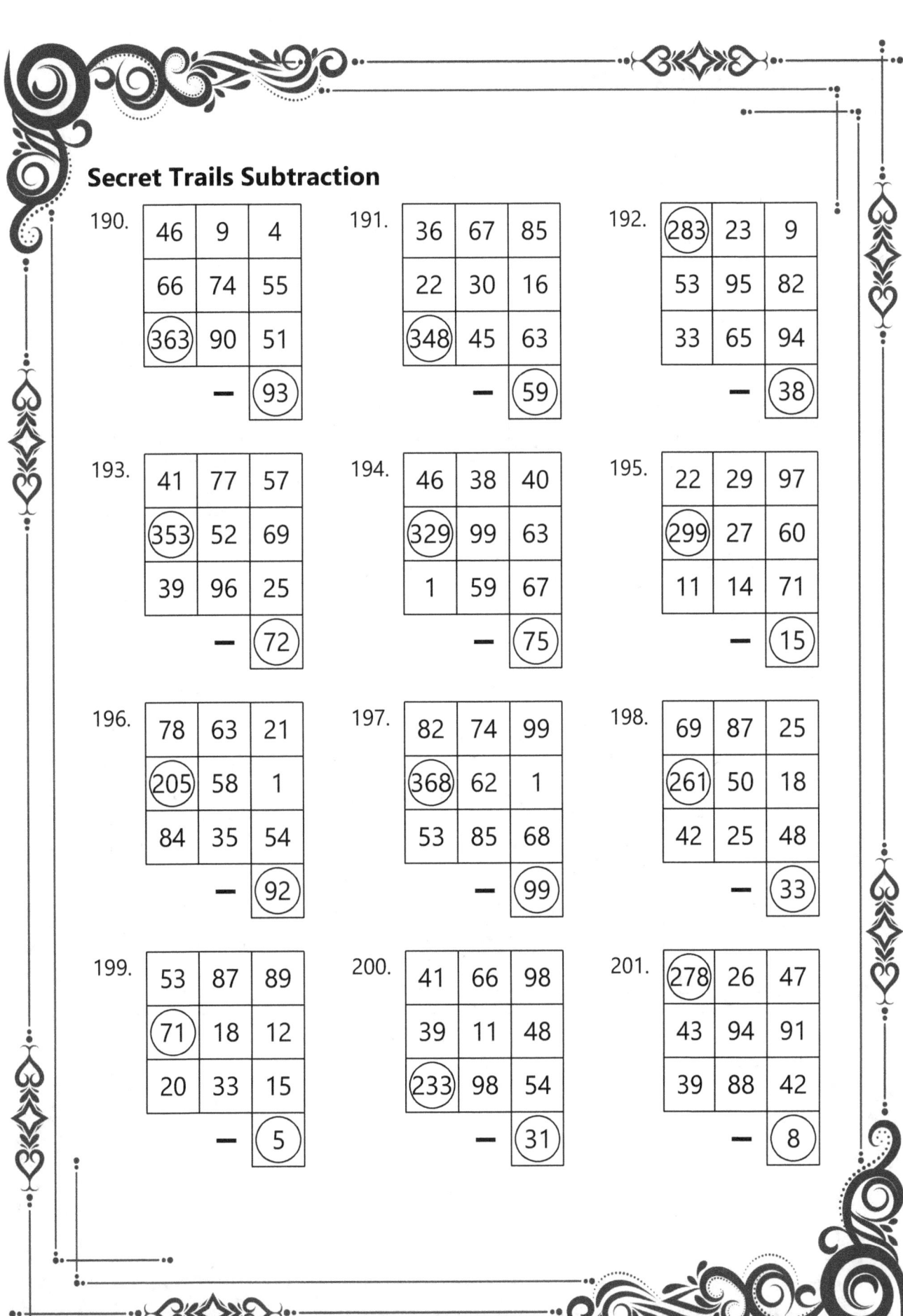

190.

46	9	4
66	74	55
(363)	90	51
	−	(93)

191.

36	67	85
22	30	16
(348)	45	63
	−	(59)

192.

(283)	23	9
53	95	82
33	65	94
	−	(38)

193.

41	77	57
(353)	52	69
39	96	25
	−	(72)

194.

46	38	40
(329)	99	63
1	59	67
	−	(75)

195.

22	29	97
(299)	27	60
11	14	71
	−	(15)

196.

78	63	21
(205)	58	1
84	35	54
	−	(92)

197.

82	74	99
(368)	62	1
53	85	68
	−	(99)

198.

69	87	25
(261)	50	18
42	25	48
	−	(33)

199.

53	87	89
(71)	18	12
20	33	15
	−	(5)

200.

41	66	98
39	11	48
(233)	98	54
	−	(31)

201.

(278)	26	47
43	94	91
39	88	42
	−	(8)

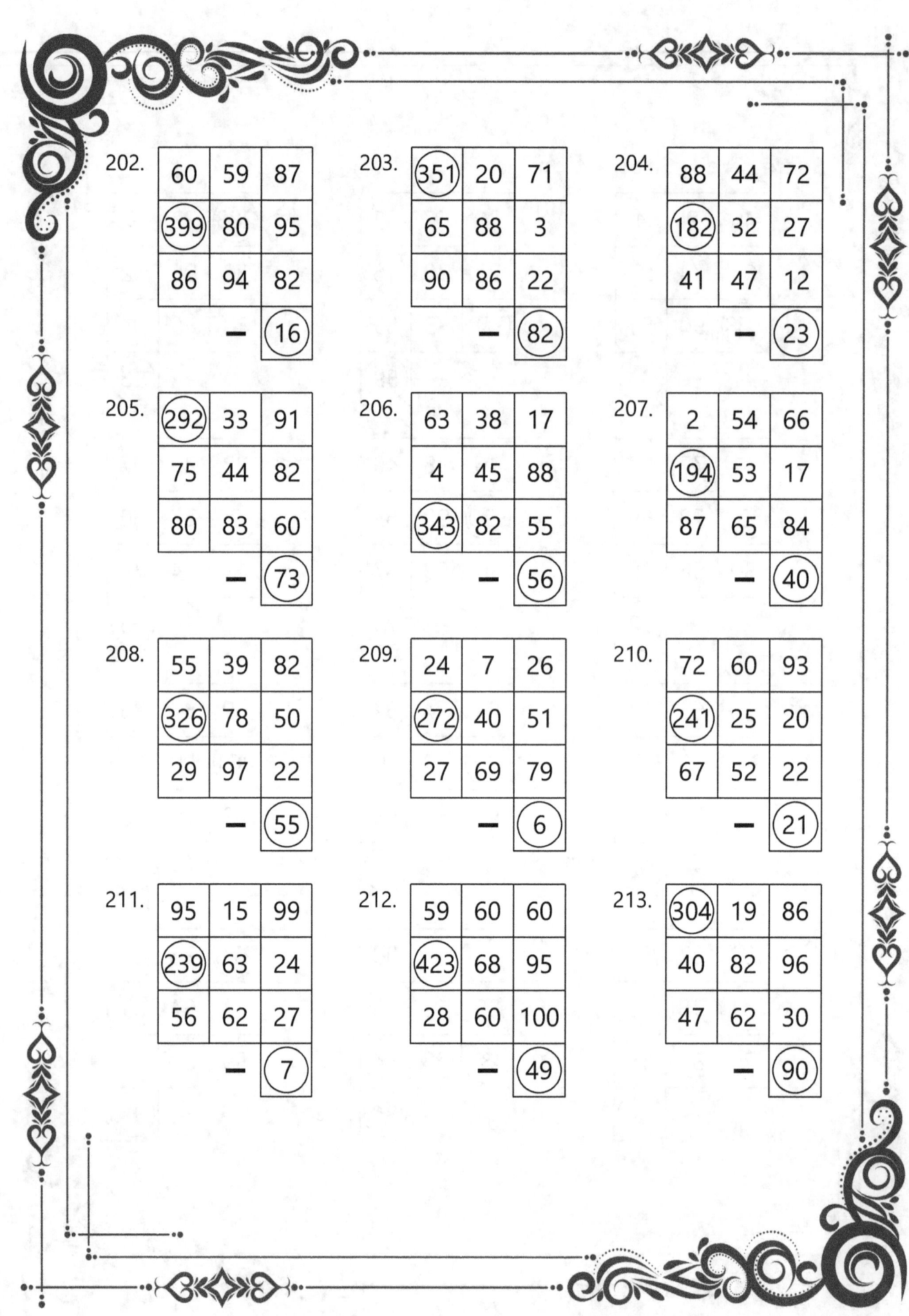

202.

60	59	87
(399)	80	95
86	94	82
	—	(16)

203.

(351)	20	71
65	88	3
90	86	22
	—	(82)

204.

88	44	72
(182)	32	27
41	47	12
	—	(23)

205.

(292)	33	91
75	44	82
80	83	60
	—	(73)

206.

63	38	17
4	45	88
(343)	82	55
	—	(56)

207.

2	54	66
(194)	53	17
87	65	84
	—	(40)

208.

55	39	82
(326)	78	50
29	97	22
	—	(55)

209.

24	7	26
(272)	40	51
27	69	79
	—	(6)

210.

72	60	93
(241)	25	20
67	52	22
	—	(21)

211.

95	15	99
(239)	63	24
56	62	27
	—	(7)

212.

59	60	60
(423)	68	95
28	60	100
	—	(49)

213.

(304)	19	86
40	82	96
47	62	30
	—	(90)

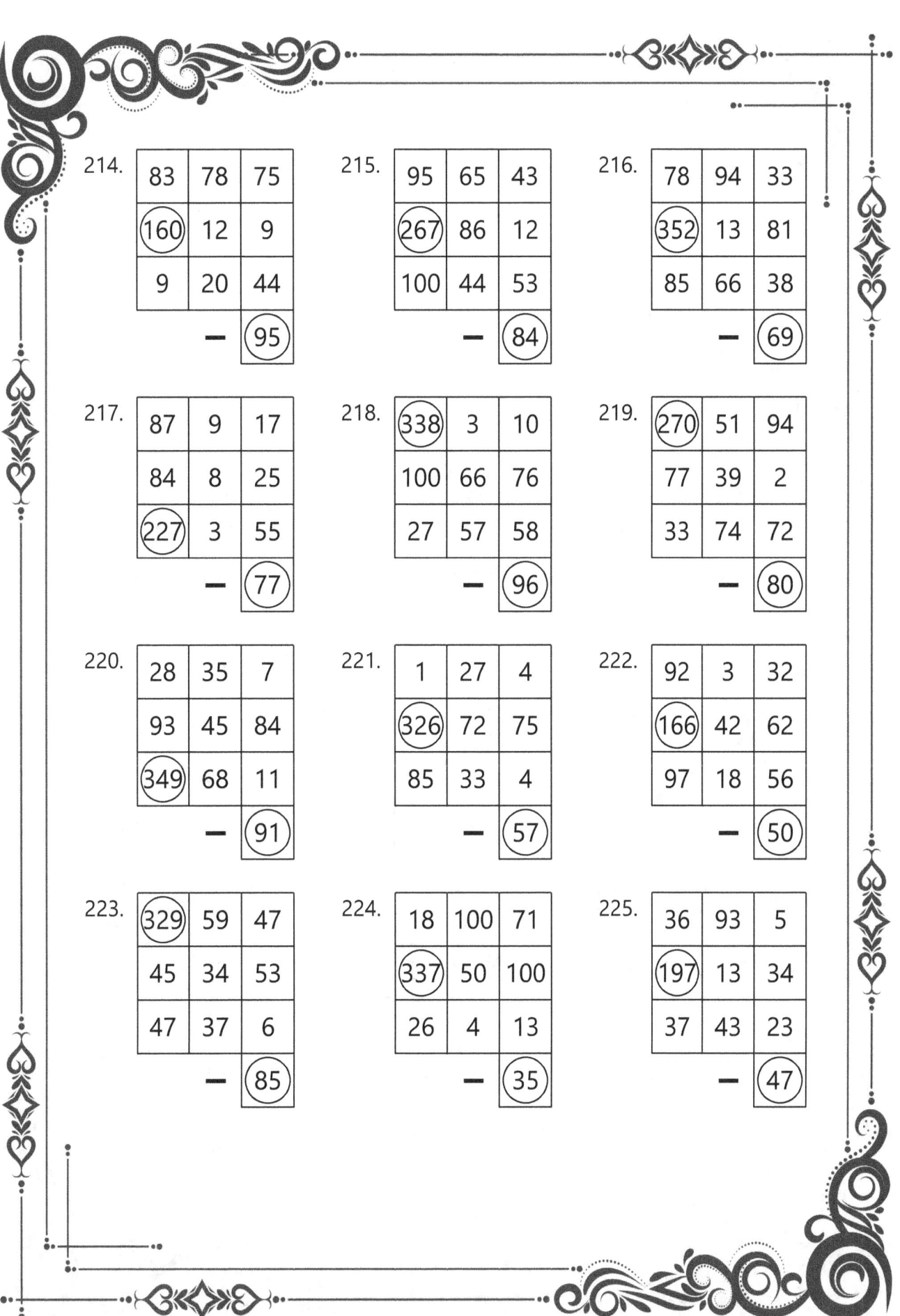

214.
83	78	75
(160)	12	9
9	20	44
	—	(95)

215.
95	65	43
(267)	86	12
100	44	53
	—	(84)

216.
78	94	33
(352)	13	81
85	66	38
	—	(69)

217.
87	9	17
84	8	25
(227)	3	55
	—	(77)

218.
(338)	3	10
100	66	76
27	57	58
	—	(96)

219.
(270)	51	94
77	39	2
33	74	72
	—	(80)

220.
28	35	7
93	45	84
(349)	68	11
	—	(91)

221.
1	27	4
(326)	72	75
85	33	4
	—	(57)

222.
92	3	32
(166)	42	62
97	18	56
	—	(50)

223.
(329)	59	47
45	34	53
47	37	6
	—	(85)

224.
18	100	71
(337)	50	100
26	4	13
	—	(35)

225.
36	93	5
(197)	13	34
37	43	23
	—	(47)

226.

58	93	22
97	86	51
(487)	48	83
	−	(22)

227.

18	62	35
85	19	91
(184)	41	15
	−	(18)

228.

60	40	42
(300)	71	98
57	38	97
	−	(34)

229.

17	96	29
(239)	71	58
39	50	24
	−	(94)

230.

89	91	38
(356)	95	79
21	2	14
	−	(65)

231.

28	42	4
(278)	3	64
11	48	63
	−	(89)

232.

59	8	15
(319)	51	80
48	50	53
	−	(98)

233.

73	32	3
(346)	73	47
88	50	63
	−	(25)

234.

(233)	74	39
1	64	27
67	47	25
	−	(68)

235.

34	36	55
(350)	46	70
35	98	100
	−	(1)

236.

13	58	13
(238)	40	27
5	90	26
	−	(74)

237.

93	98	88
23	79	81
(246)	6	43
	−	(20)

Sudoku Easy Level

Fill the grid so that every row, every column and every 3x3 box contains the numbers 1 to 9.

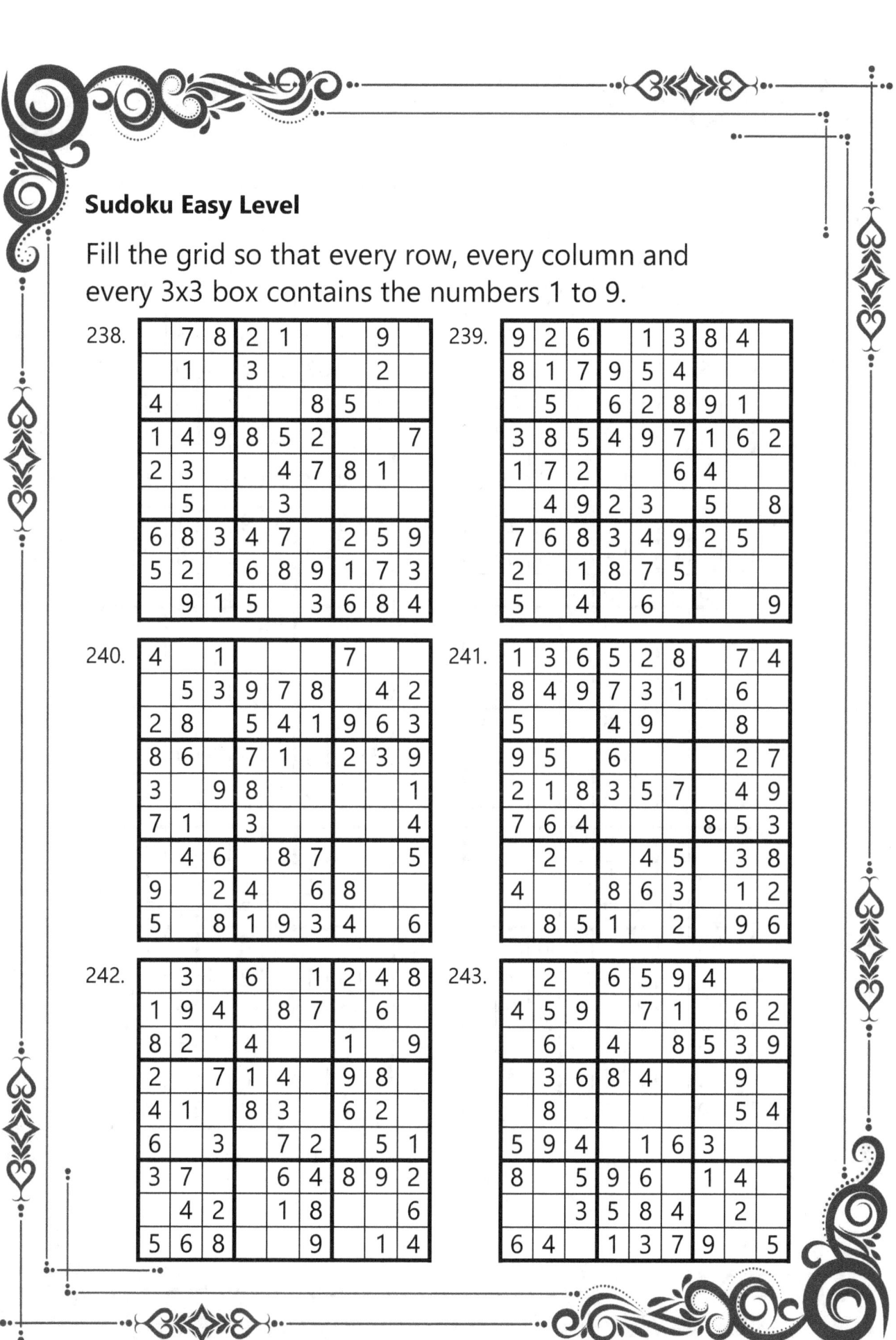

238.

239.

240.

241.

242.

243.

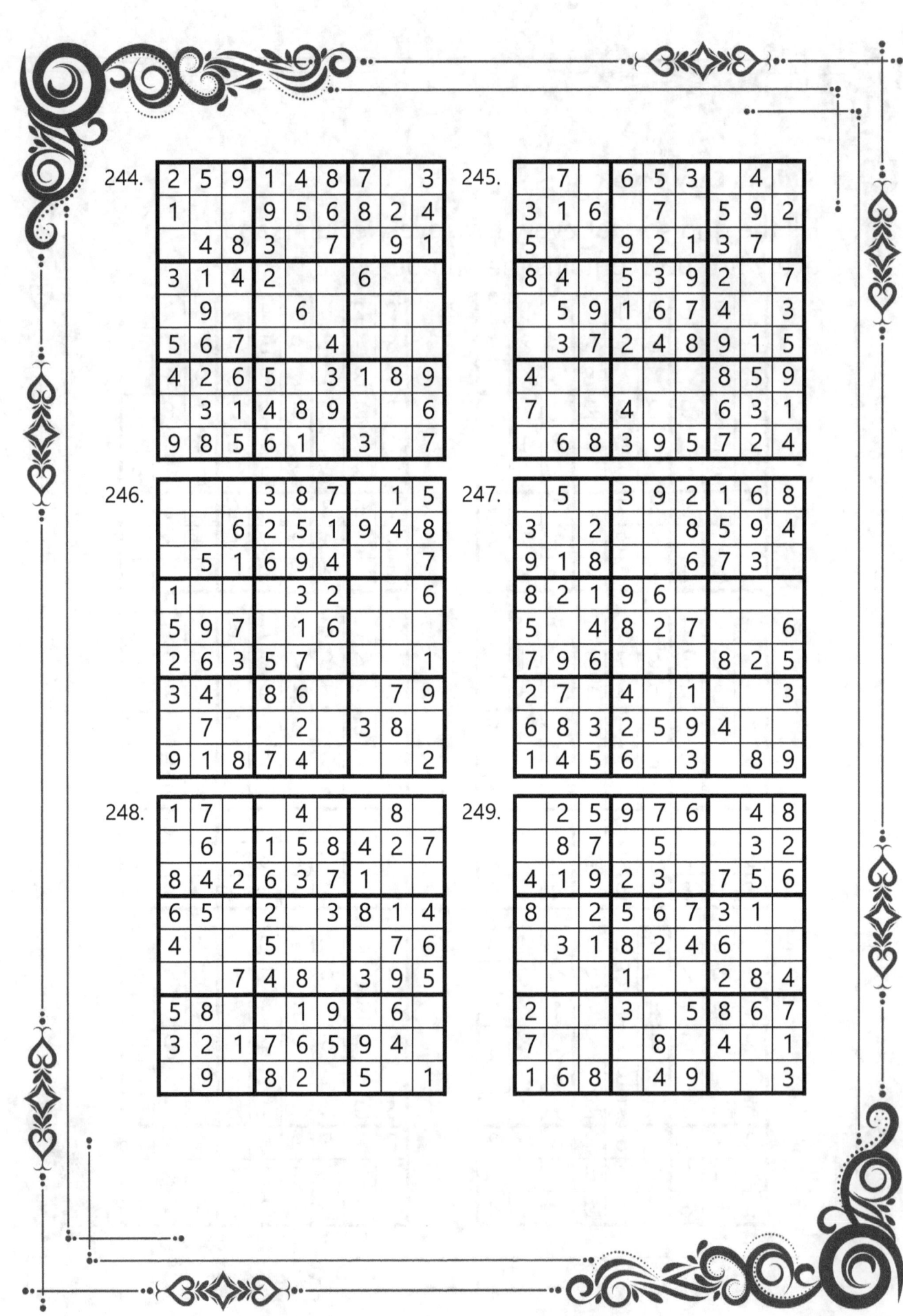

244.

2	5	9	1	4	8	7		3
1			9	5	6	8	2	4
	4	8	3		7		9	1
3	1	4	2			6		
	9			6				
5	6	7			4			
4	2	6	5		3	1	8	9
	3	1	4	8	9			6
9	8	5	6	1		3		7

245.

	7		6	5	3		4	
3	1	6		7		5	9	2
5			9	2	1	3	7	
8	4		5	3	9	2		7
	5	9	1	6	7	4		3
	3	7	2	4	8	9	1	5
4						8	5	9
7			4			6	3	1
	6	8	3	9	5	7	2	4

246.

			3	8	7		1	5
		6	2	5	1	9	4	8
	5	1	6	9	4			7
1			3	2				6
5	9	7		1	6			
2	6	3	5	7				1
3	4		8	6			7	9
	7			2		3	8	
9	1	8	7	4				2

247.

	5		3	9	2	1	6	8
3		2			8	5	9	4
9	1	8				6	7	3
8	2	1	9	6				
5		4	8	2	7			6
7	9	6				8	2	5
2	7		4		1			3
6	8	3	2	5	9	4		
1	4	5	6		3		8	9

248.

1	7			4			8	
	6		1	5	8	4	2	7
8	4	2	6	3	7	1		
6	5		2		3	8	1	4
4			5				7	6
		7	4	8		3	9	5
5	8			1	9		6	
3	2	1	7	6	5	9	4	
	9		8	2		5		1

249.

	2	5	9	7	6		4	8
	8	7		5			3	2
4	1	9	2	3		7	5	6
8		2	5	6	7	3	1	
	3	1	8	2	4	6		
			1			2	8	4
2			3		5	8	6	7
7				8		4		1
1	6	8		4	9			3

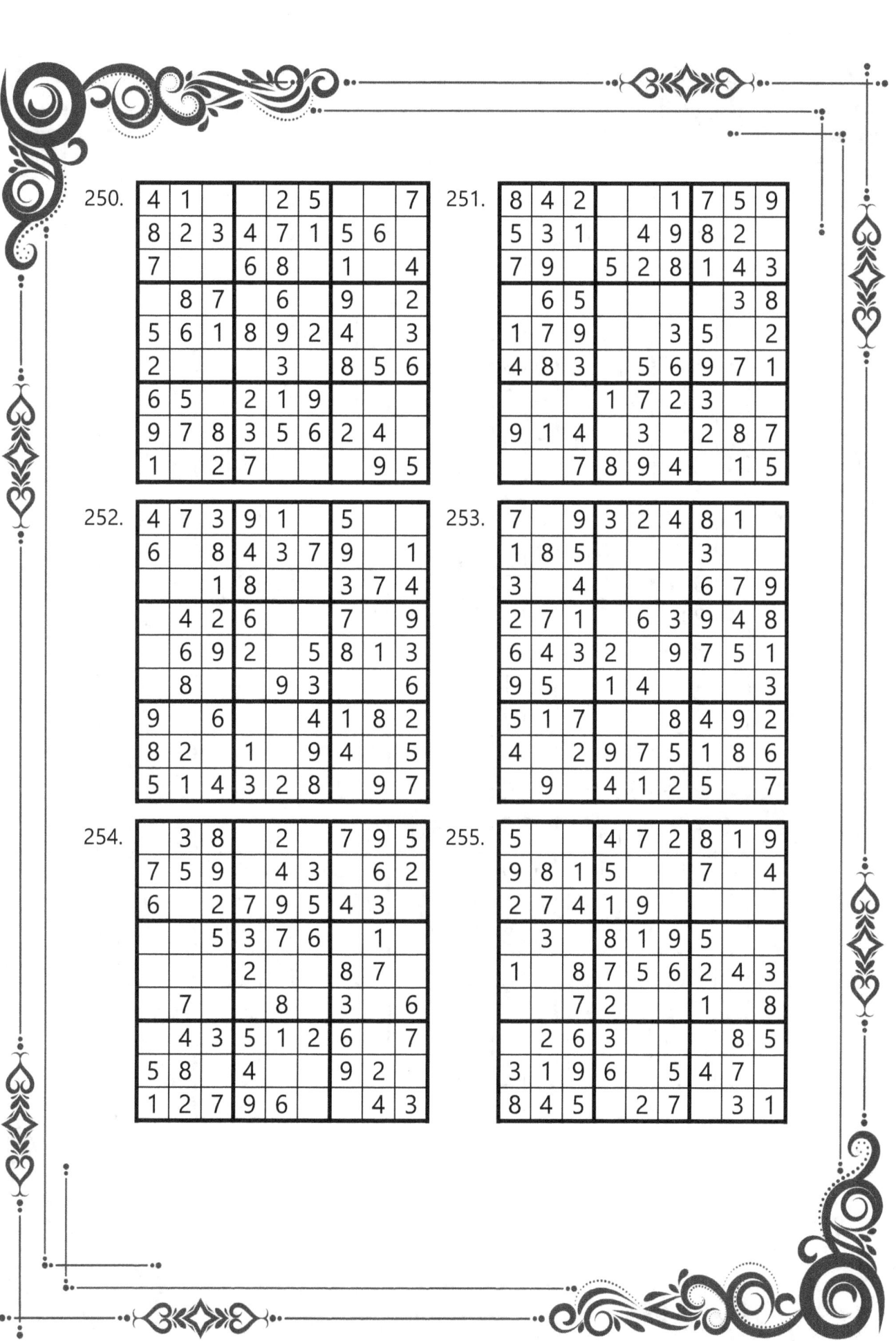

256.

	3		7	2	8	6		4
7	1	6	3			5		
8	4	2	5	6	1		9	3
1	8	9	4	3		2		
2	7	3		8			4	
5	6	4	2		9	8	3	7
		1	8		6	4	7	2
		7	1	4	2	3		9
4	2			7	3	1	5	6

257.

	6	3	1	4	2	8		9
5			9	8			6	
	9	8				1	4	3
8	7		3	1			9	
1		9	2	7	4	6	3	8
	3	4	8	5	9	7		
3	8					9	7	
4		5			7		8	1
9			4	3	8	5	2	6

258.

		4	3	7	8	2		1
	1			6		4		8
2	3	8	4	1		9	6	7
1	2		7	9	3	8		5
4	8	5	1		6	3	7	
3			5		4		1	2
	4		8		9	5	2	6
6	9	3	2	5		1	8	4
8	5	2	6	4			9	3

259.

1	5	9	3			2	6	4
7			8		9	4	1	3
	2	4	6	8	1	9	7	
5	4	6	8	2	7	3	1	
8	7	3			6		5	4
2	9		4	3		6		
	8	7	2	4	3		9	1
4			1	6	9	7		
	1		7		8	4	3	6

260.

	9	2	5	4			7	3
8				1	2	6	9	5
5	6	7	9	3	8	2		
4			2		5	7	3	6
6	5	3	4	8		1		9
7	2	9		6		4		8
9	7	1	8		4	3	6	2
	8		3		9	5		4
3				2	1	9	8	7

261.

3		7	8			6	4	5
5	8	1		3	4	7		9
9		6		2	7		1	8
8			1		3	4	5	2
4	3	5	9	8	2			6
	7	2	4	5	6	8	9	3
6		3	2		9	5	8	
2				6		9		
			3	1	5		6	4

262.

7		6	8	1	4		3	
2	1		6	3			4	8
	3	4	7	2	5	9	6	
3		2	1			4	9	
1				3				
	6	9	2	5	7			3
	2			4	6	8	1	
			5	8	1	6	2	7
6	8	1		7	2	3		4

263.

8	3	1	9	6	5	4		7
	6	2	3	4	7		9	8
9			1	2		6		5
4	5	6	8	7		2	1	9
			4	1	2		6	3
					9		7	
3		5	2		6	7	4	1
2	1			8		3	5	6
6	7	4	5		1		8	2

264.

		5			2	6	1	3
	1		5	4	6		8	7
	6			8	3	2	4	5
9		7	3	1			2	
		8	9	5	7	4	3	1
1	3	4	2				5	
		6	4		5	1		2
5	7		8	2	9	3	6	4
2	4						9	8

265.

4	5	7				6	1	
3		6		1	4		8	2
	1	8		9	5	4	3	7
1		4	9	5	2	7	6	3
7	3	2	1		8	9	4	5
	6	9	3			8	2	1
9	2	5	4			1		8
8				7	1	2	9	
	7		2	8		3	5	4

266.

8	1	3	7	4				2
7		6	8	3			1	
		2	6	9			7	8
9	7		2	1	3	5	4	
	2	5		6	7		3	1
6	3		4		8		2	9
3				2	9	1	6	7
1		9	3	7	6	2	8	
			1	8	4	9		3

267.

2	5	7	8		6		9	3
	9	1	3	5		7		2
			9		7		1	
4	6				5	3	2	
7	2		1	4		9	6	5
1	3	5		6		8	4	7
8	1	3			2	6		
5		2	6	7		1		9
		6			1		5	8

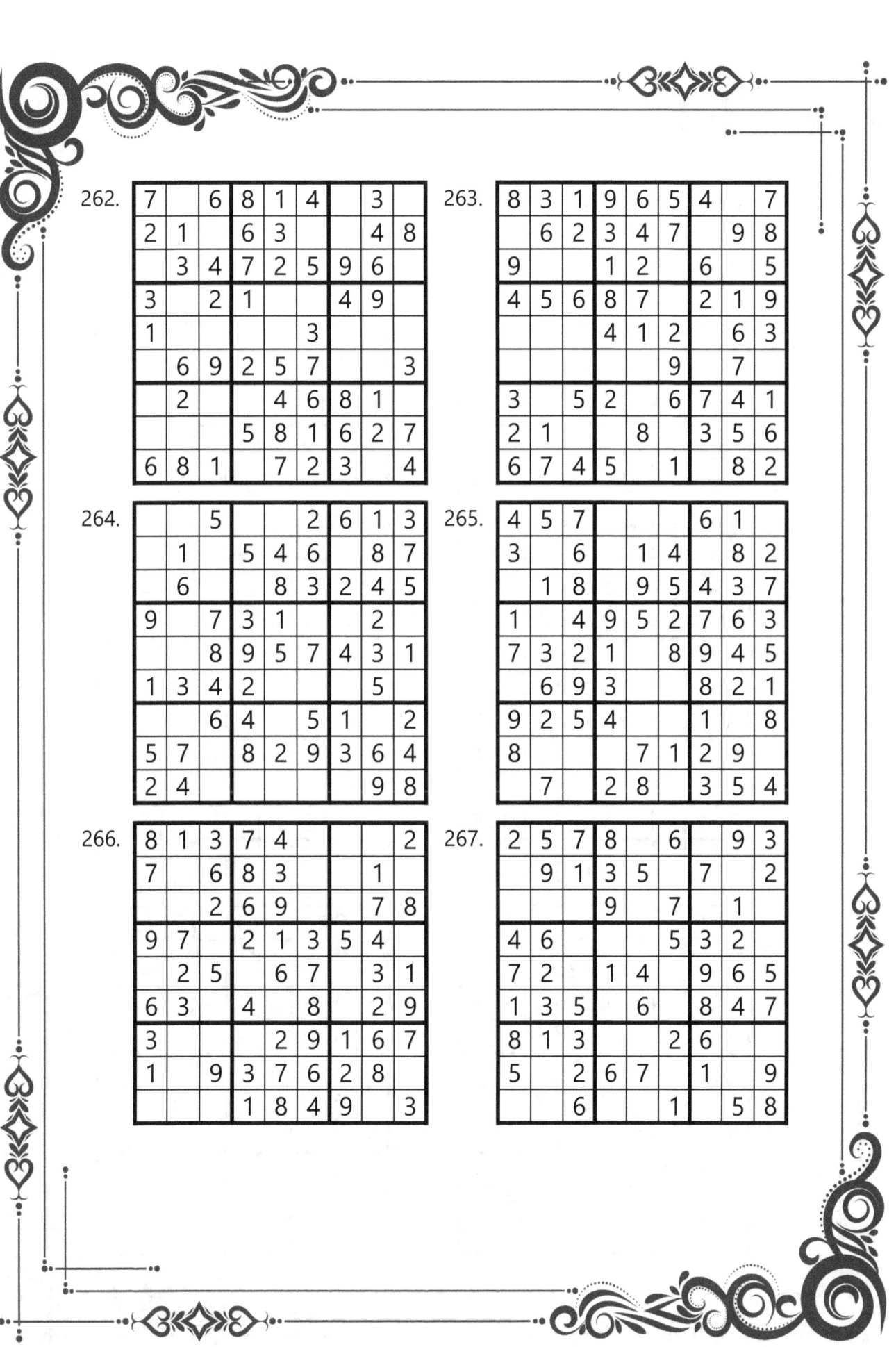

Sudoku Medium Level

Fill the grid so that every row, every column and every 3x3 box contains the numbers 1 to 9.

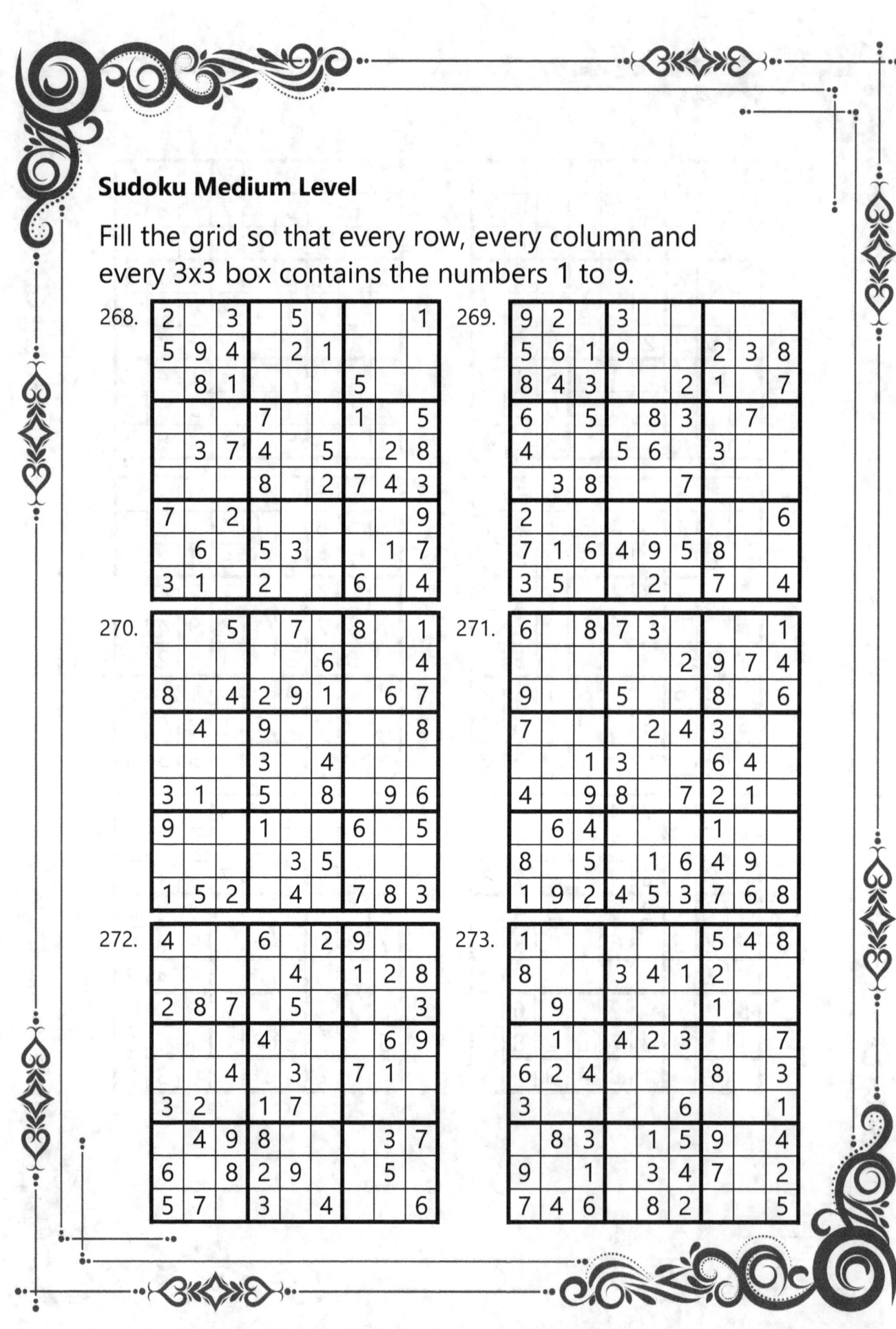

268.

2		3		5				1
5	9	4			2	1		
	8	1				5		
			7			1		5
	3	7	4		5		2	8
			8		2	7	4	3
7		2						9
	6			5	3		1	7
3	1			2			6		4

269.

9	2		3					
5	6	1	9			2	3	8
8	4	3				2	1		7
6		5		8	3		7	
4			5	6		3		
	3	8			7			
2								6
7	1	6	4	9	5	8		
3	5			2			7		4

270.

		5		7		8		1
				6				4
8		4	2	9	1		6	7
	4		9					8
			3		4			
3	1		5		8		9	6
9			1			6		5
				3	5			
1	5	2		4		7	8	3

271.

6		8	7	3				1
				2	9	7	4
9			5			8		6
7				2	4	3		
		1	3			6	4	
4		9	8		7	2	1	
	6	4			1		
8		5		1	6	4	9	
1	9	2	4	5	3	7	6	8

272.

4			6		2	9		
			4		1	2	8
2	8	7		5				3
			4				6	9
		4		3		7	1
3	2		1	7			
	4	9	8			3	7
6		8	2	9		5	
5	7		3		4		6

273.

1					5	4	8
8			3	4	1	2	
	9					1	
	1		4	2	3			7
6	2	4				8		3
3				6				1
	8	3		1	5	9		4
9		1		3	4	7		2
7	4	6		8	2			5

274. 275.

276. 277.

278. 279.

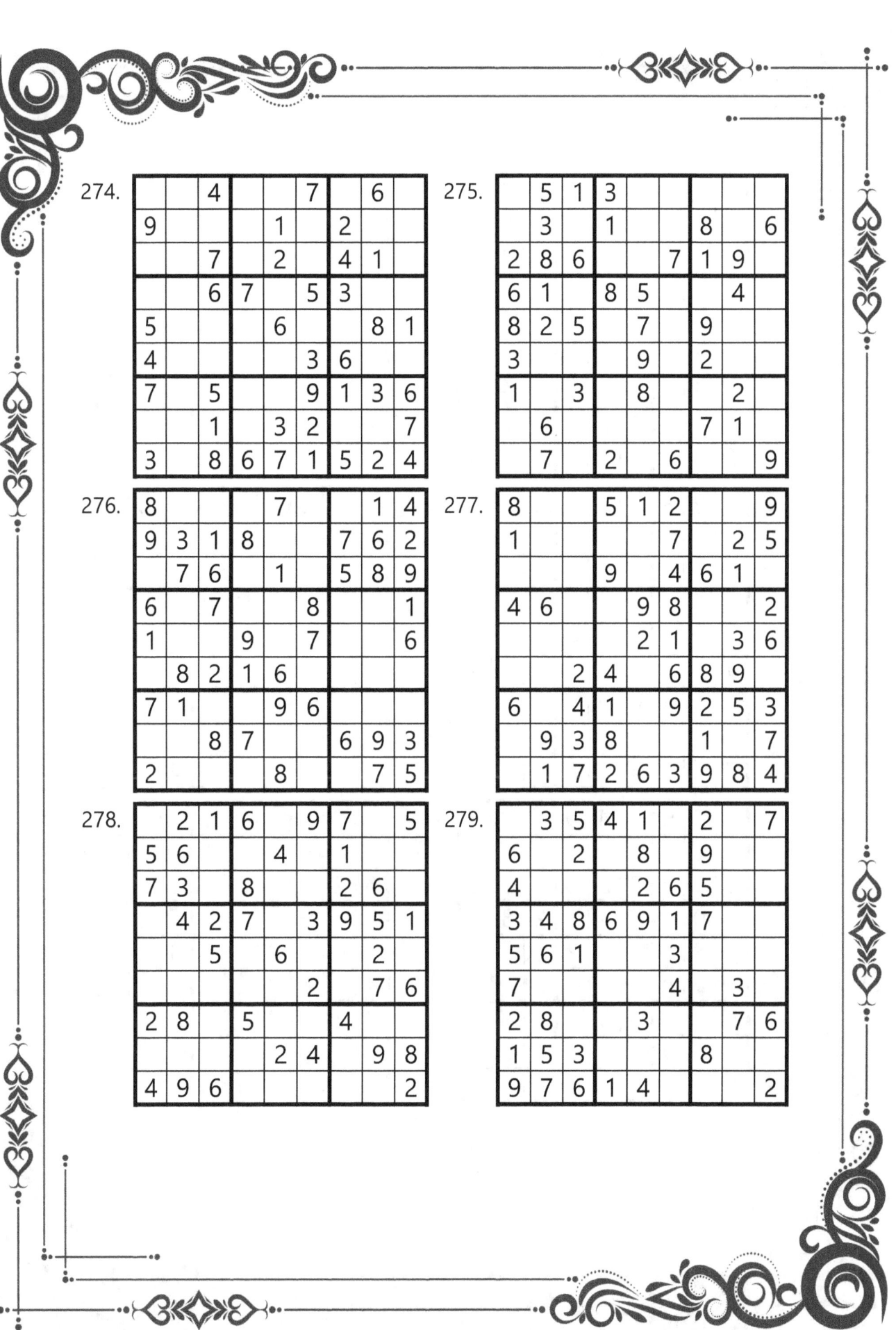

280.

	4					2	7	3
3						7	2	5
1	2	7			5	6	4	8
	8		1	7	4	9		
2		6		8	9	4	1	5
4	1		2			3	8	
7	5				1		9	
	9		5	2			7	
6				9	8			4

281.

		1				4		7
4			1		6		5	
3		9	5	7	4	8	6	1
	3	2	4			7	6	
9				1	8		2	4
7	4	8			9	3	1	
6			7	9				8
2			3	8	4		7	
	7	4	2			1	9	3

282.

	2	3	5			9		
			7					
9		7	1	3		6		
3	9	2						
6	7		2		4	3	1	
4		1	3					2
	3	4	6	2				
	1			7				4
7			4		8		3	1

283.

	9	5	6	4			8	3
6			8			2		4
			1			7	6	
1	2	7		4	8		9	
9	5	3	2	1	6			7
			7	9		3		
3	4		9	5				8
7			6	4		3		
	9	8				4	3	1

284.

3	2	6		4	9			5
9						4	7	2
7		1	5	2	8		9	3
8	9		4		6	2	3	
	5				2			7
					3	5		
			6	3		9		4
		9		1		7		
5						7	3	6

285.

	9	8	4	2	1			6
		3	4	9				2
6	2			8		1	9	4
2		9	1				4	
	6		2	5		8	1	
4			1	7	3			
9	1		3			7	2	6
						4	5	1
		2		1	5	9	7	3

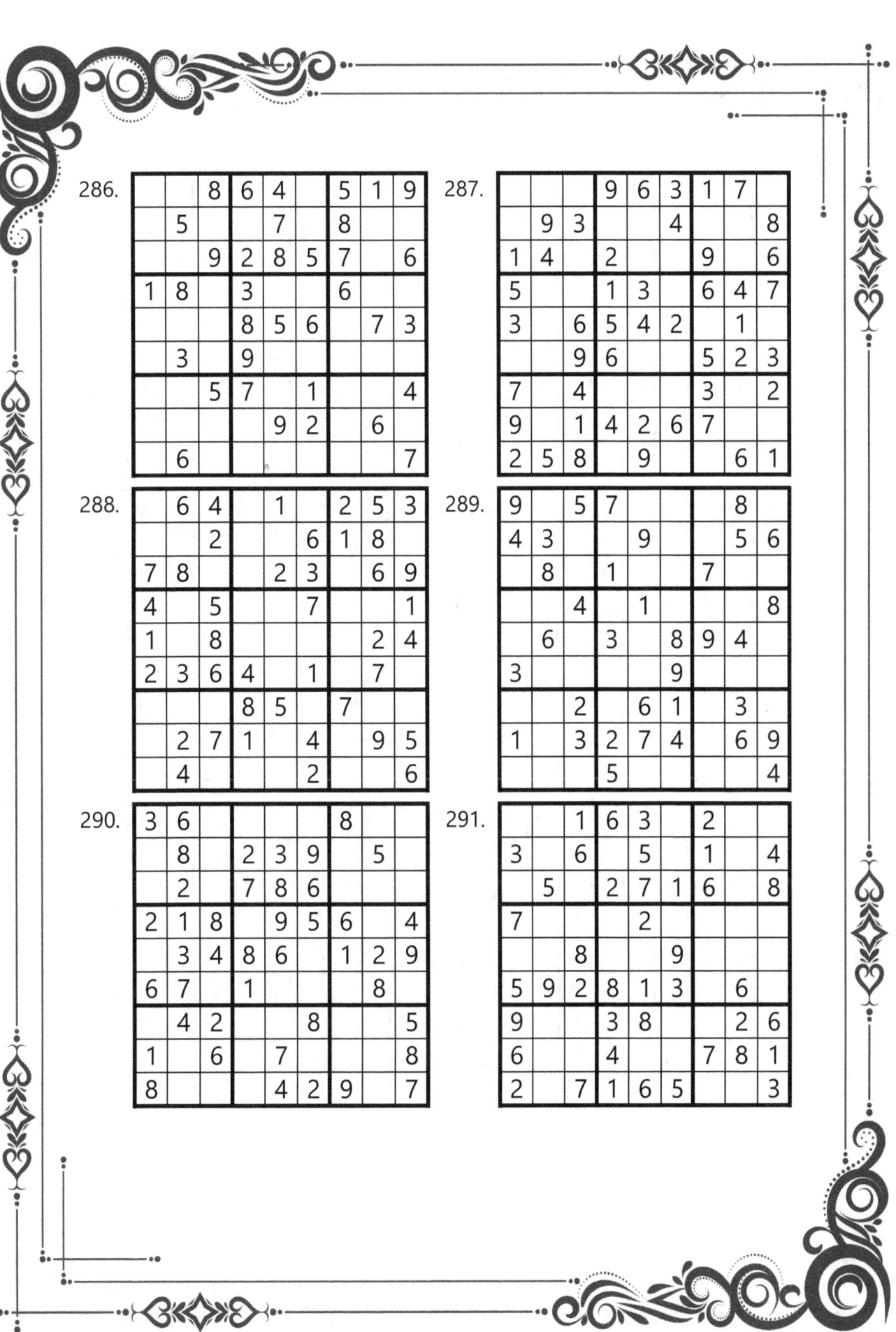

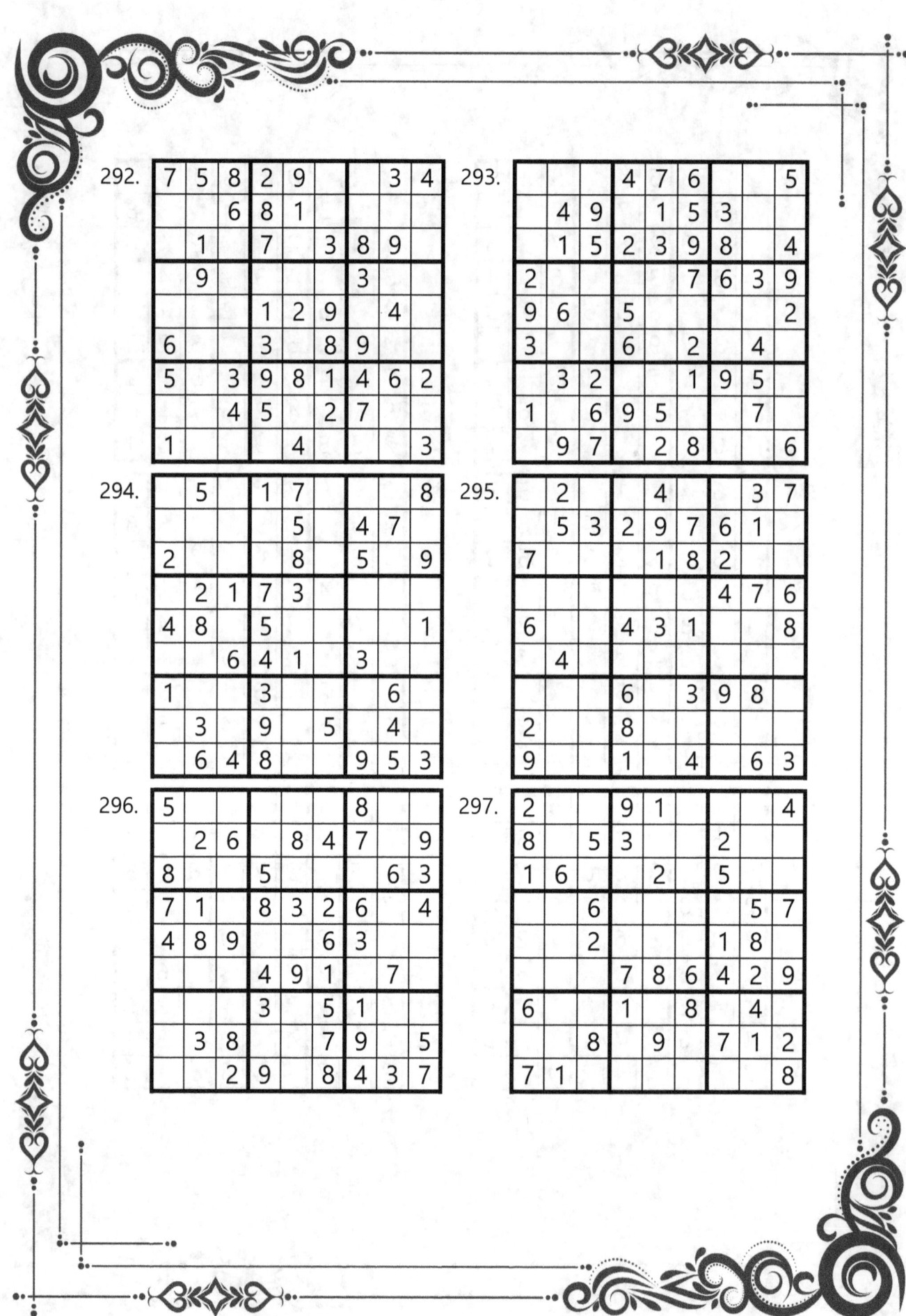

292.

7	5	8	2	9			3	4
		6	8	1				
	1		7		3	8	9	
	9					3		
			1	2	9		4	
6			3		8	9		
5		3	9	8	1	4	6	2
		4	5		2	7		
1				4				3

293.

			4	7	6			5
	4	9		1	5	3		
	1	5	2	3	9	8		4
2					7	6	3	9
9	6		5					2
3			6		2		4	
	3	2				1	9	5
1			6	9	5		7	
	9	7		2	8			6

294.

	5		1	7				8
				5		4	7	
2				8		5		9
	2	1	7	3				
4	8		5					1
		6	4	1		3		
1			3			6		
	3		9		5		4	
	6	4	8			9	5	3

295.

	2			4			3	7
	5	3	2	9	7	6	1	
7				1	8	2		
						4	7	6
6			4	3	1			8
	4							
			6		3	9	8	
2			8					
9			1		4		6	3

296.

5						8		
	2	6		8	4	7		9
8			5				6	3
7	1		8	3	2	6		4
4	8	9			6	3		
			4	9	1		7	
			3		5	1		
	3	8			7	9		5
		2	9		8	4	3	7

297.

2			9	1				4
8			5	3			2	
1	6			2			5	
			6				5	7
			2				1	8
			7	8	6	4	2	9
6				1		8		4
		8		9		7	1	2
7	1							8

Sudoku Hard Level

Fill the grid so that every row, every column and every 3x3 box contains the numbers 1 to 9.

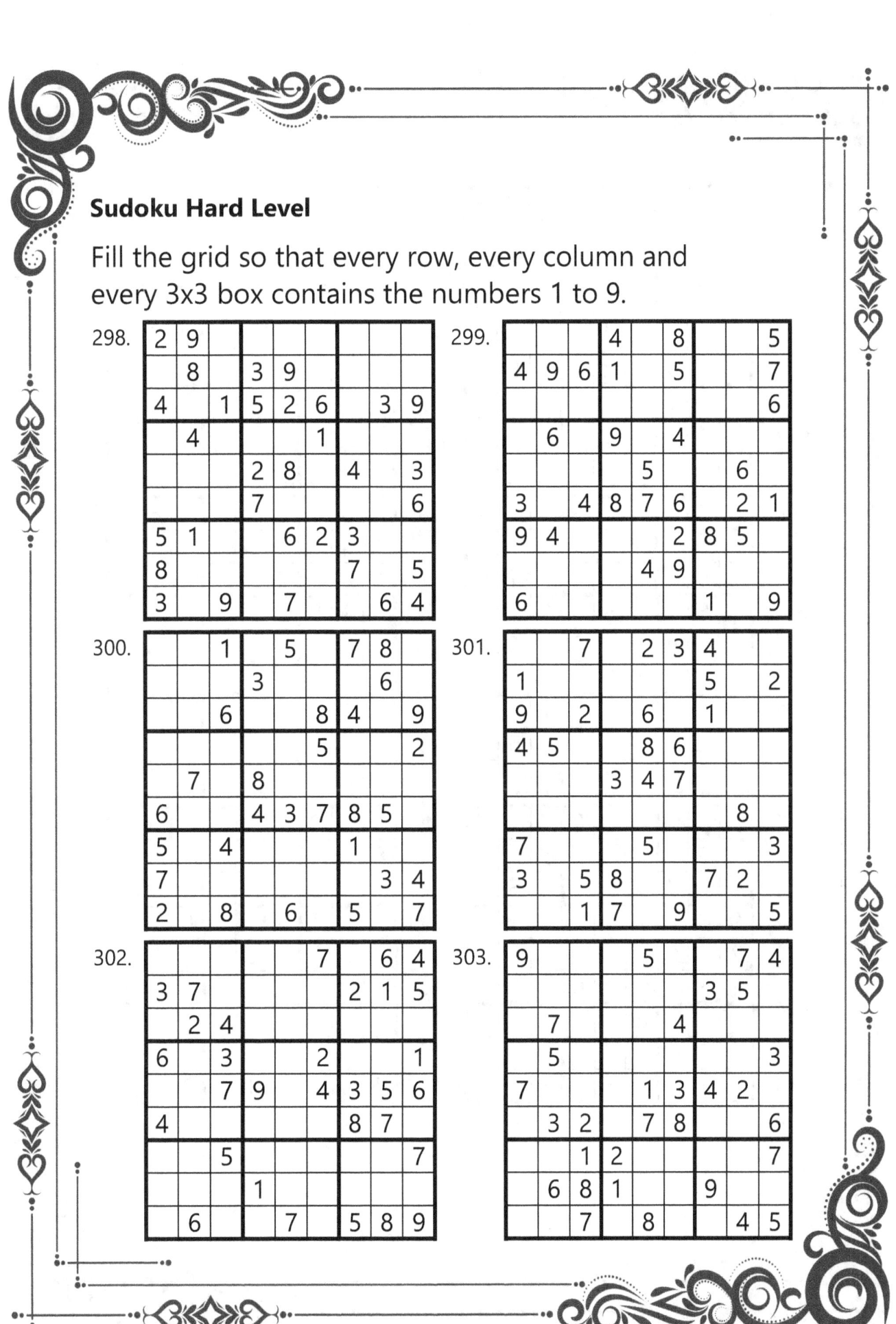

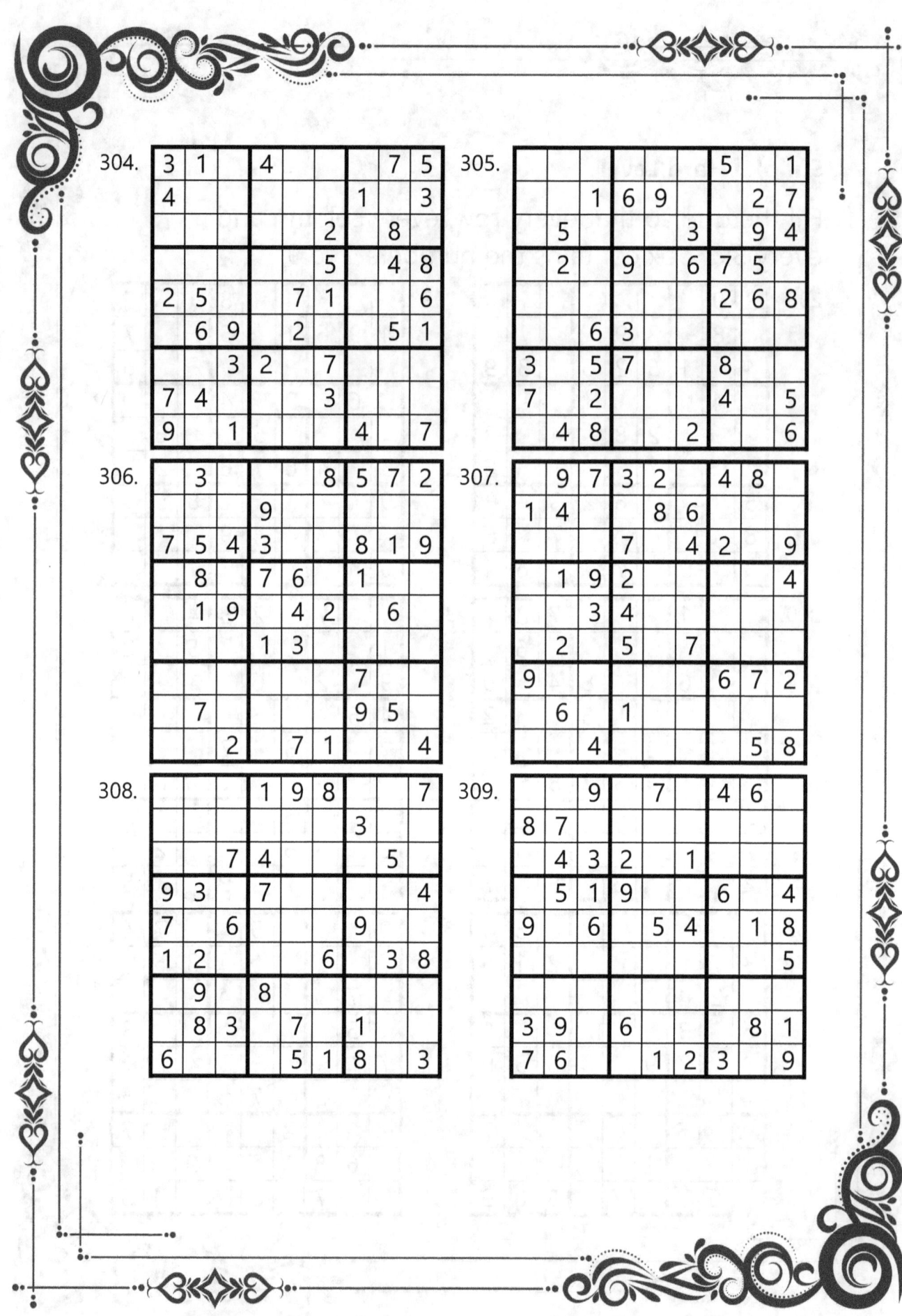

304. 305. 306. 307. 308. 309.

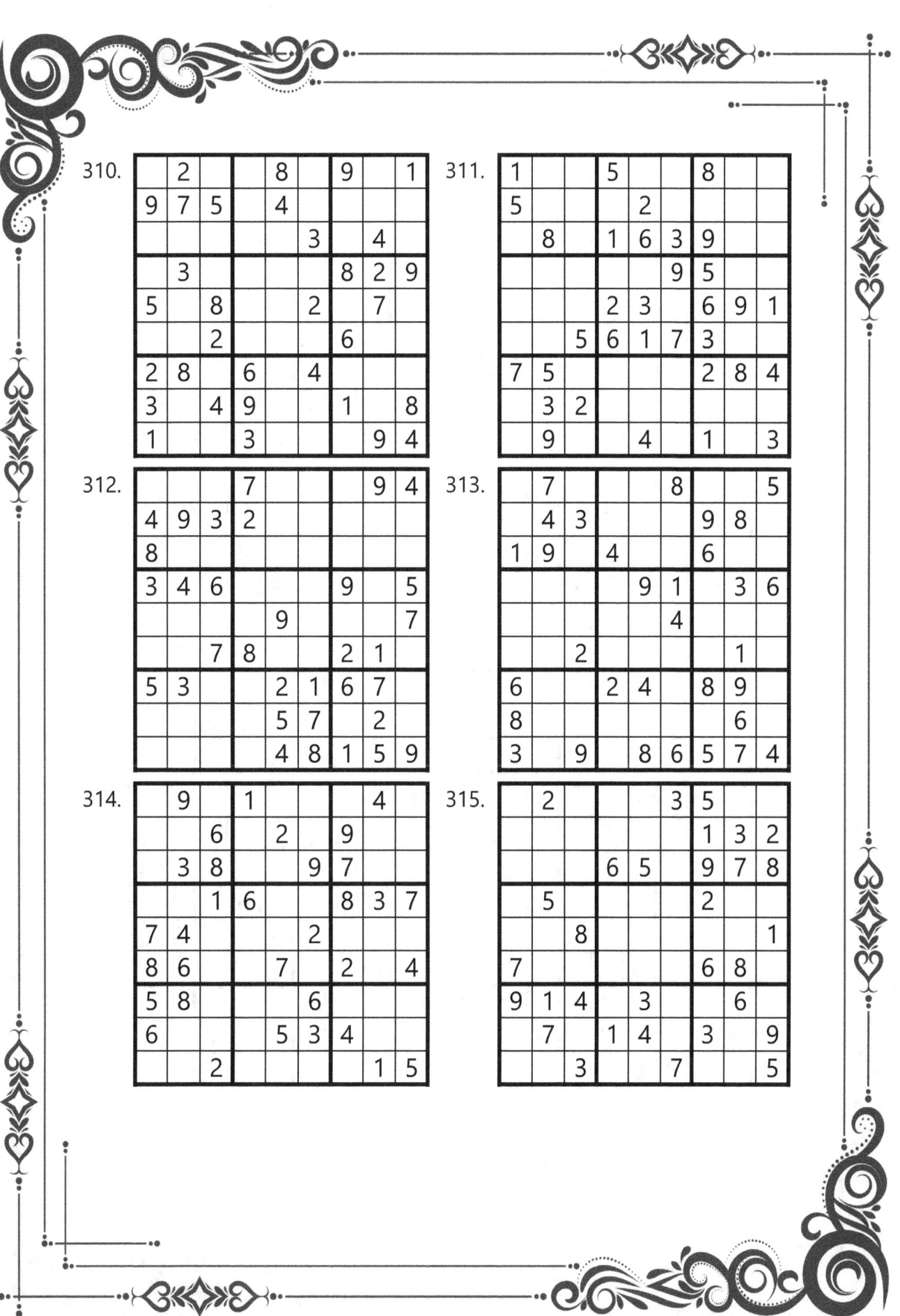

310.

	2			8		9		1
9	7	5		4				
					3		4	
	3					8	2	9
5		8			2		7	
		2				6		
2	8		6		4			
3		4	9			1		8
1			3			9	4	

311.

1			5			8		
5				2				
	8		1	6	3	9		
						9	5	
			2	3		6	9	1
		5	6	1	7	3		
7	5					2	8	4
	3	2						
	9			4		1		3

312.

			7				9	4
4	9	3	2					
8								
3	4	6				9		5
				9				7
		7	8			2	1	
5	3			2	1	6	7	
				5	7		2	
				4	8	1	5	9

313.

	7				8			5
	4	3				9	8	
1	9		4			6		
				9	1		3	6
					4			
		2					1	
6			2	4		8	9	
8							6	
3		9		8	6	5	7	4

314.

	9		1				4	
		6		2		9		
	3	8			9	7		
		1	6			8	3	7
7	4				2			
8	6			7		2		4
5	8				6			
6				5	3	4		
		2					1	5

315.

	2					3	5	
						1	3	2
			6	5		9	7	8
	5					2		
		8						1
7						6	8	
9	1	4		3			6	
	7		1	4		3		9
		3			7			5

316.

				9		1		
	5	4		8			2	
		2		5	4		3	
6						4		
4		8		2				
				4	2		1	
	7	3	8		6	5		9
		1			6		4	
9		4	1	5	7			8

317.

								9
		1			4	7		
			2				3	
1		7	5	8				4
		3	6	4		5	9	
4				9			8	1
3						6	4	
6		4					2	
5			4			8	9	3

318.

						6		
	1		9	6	4	7		2
7				5				
9		7	5	3	2			
				2				
		2	8		1		4	7
	7		2	8		4		
8					6			1
6						3	2	8

319.

3			5	1				
9								4
	7		9	4	6			
7		3	2	9		1		
	1				3	4	9	
6				8		3	5	7
				8		2	5	
		7						
5		2	4					1

320.

6			3	9	1	4		5
3				6		2	1	
			2	5				
	4	8		7		6		
			9					4
			5	3	4		2	
		7	6	2	9			
4						5		
		3	8					2

321.

						1		
6	2	5					1	
1	9			3			4	
8	3			6		9		
	6		9			3	5	7
4			3					1
		6	2			8		
	1			4	6	7	2	
5					1			6

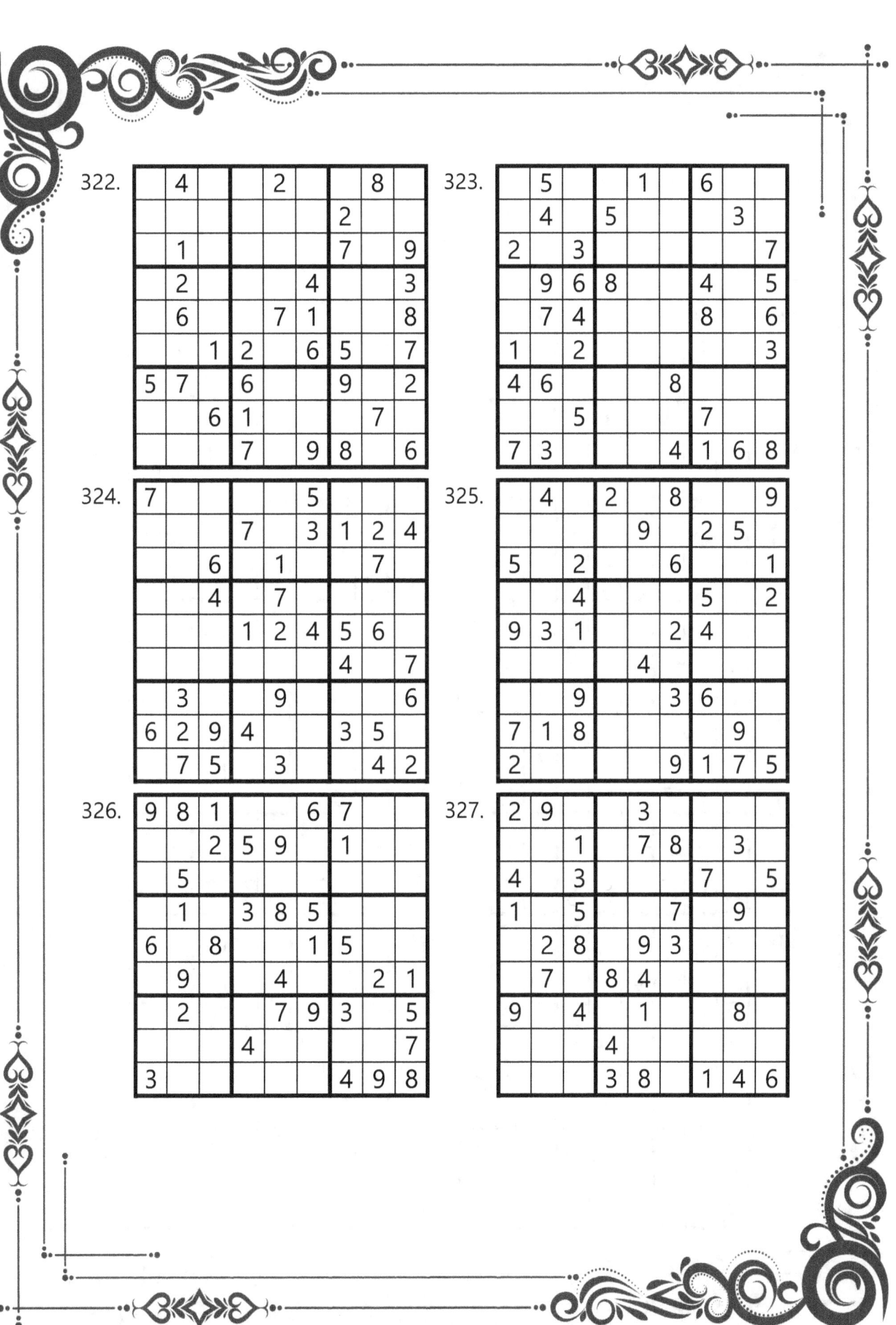

322.

```
. 4 . | . 2 . | . 8 .
. . . | . . . | 2 . .
. 1 . | . . . | 7 . 9
------+-------+------
. 2 . | . . 4 | . . 3
. 6 . | . 7 1 | . . 8
. . 1 | 2 . . | 6 5 7
------+-------+------
5 7 . | 6 . . | 9 . 2
. . 6 | 1 . . | 7 . .
. . . | 7 . . | 9 8 6
```

323.

```
. 5 . | . 1 . | 6 . .
. 4 . | 5 . . | . 3 .
2 . 3 | . . . | . . 7
------+-------+------
. 9 6 | 8 . . | 4 . 5
. 7 4 | . . . | 8 . 6
1 . 2 | . . . | . . 3
------+-------+------
4 6 . | . . 8 | . . .
. . 5 | . . . | 7 . .
7 3 . | . . 4 | 1 6 8
```

324.

```
7 . . | . 5 . | . . .
. . . | 7 . 3 | 1 2 4
. . 6 | . 1 . | . 7 .
------+-------+------
. . 4 | . 7 . | . . .
. . . | 1 2 4 | 5 6 .
. . . | . . . | 4 . 7
------+-------+------
. 3 . | . 9 . | . . 6
6 2 9 | 4 . . | 3 5 .
. 7 5 | . 3 . | . 4 2
```

325.

```
. 4 . | 2 . 8 | . . 9
. . . | . 9 . | 2 5 .
5 . 2 | . . 6 | . . 1
------+-------+------
. . 4 | . . . | 5 . 2
9 3 1 | . . 2 | 4 . .
. . . | . 4 . | . . .
------+-------+------
. . 9 | . . 3 | 6 . .
7 1 8 | . . . | . 9 .
2 . . | . . 9 | 1 7 5
```

326.

```
9 8 1 | . . 6 | 7 . .
. . 2 | 5 9 . | 1 . .
. 5 . | . . . | . . .
------+-------+------
. 1 . | 3 8 5 | . . .
6 . 8 | . . 1 | 5 . .
. 9 . | . 4 . | . 2 1
------+-------+------
. 2 . | . 7 9 | 3 . 5
. . . | 4 . . | . . 7
3 . . | . . . | 4 9 8
```

327.

```
2 9 . | . 3 . | . . .
. . 1 | . 7 8 | . 3 .
4 . 3 | . . . | 7 . 5
------+-------+------
1 . 5 | . . 7 | . 9 .
. . 2 | 8 . 9 | 3 . .
. . 7 | . 8 4 | . . .
------+-------+------
9 . 4 | . 1 . | . 8 .
. . 4 | . . . | . . .
. . . | 3 8 . | 1 4 6
```

Sudoku Expert Level

Fill the grid so that every row, every column and every 3x3 box contains the numbers 1 to 9.

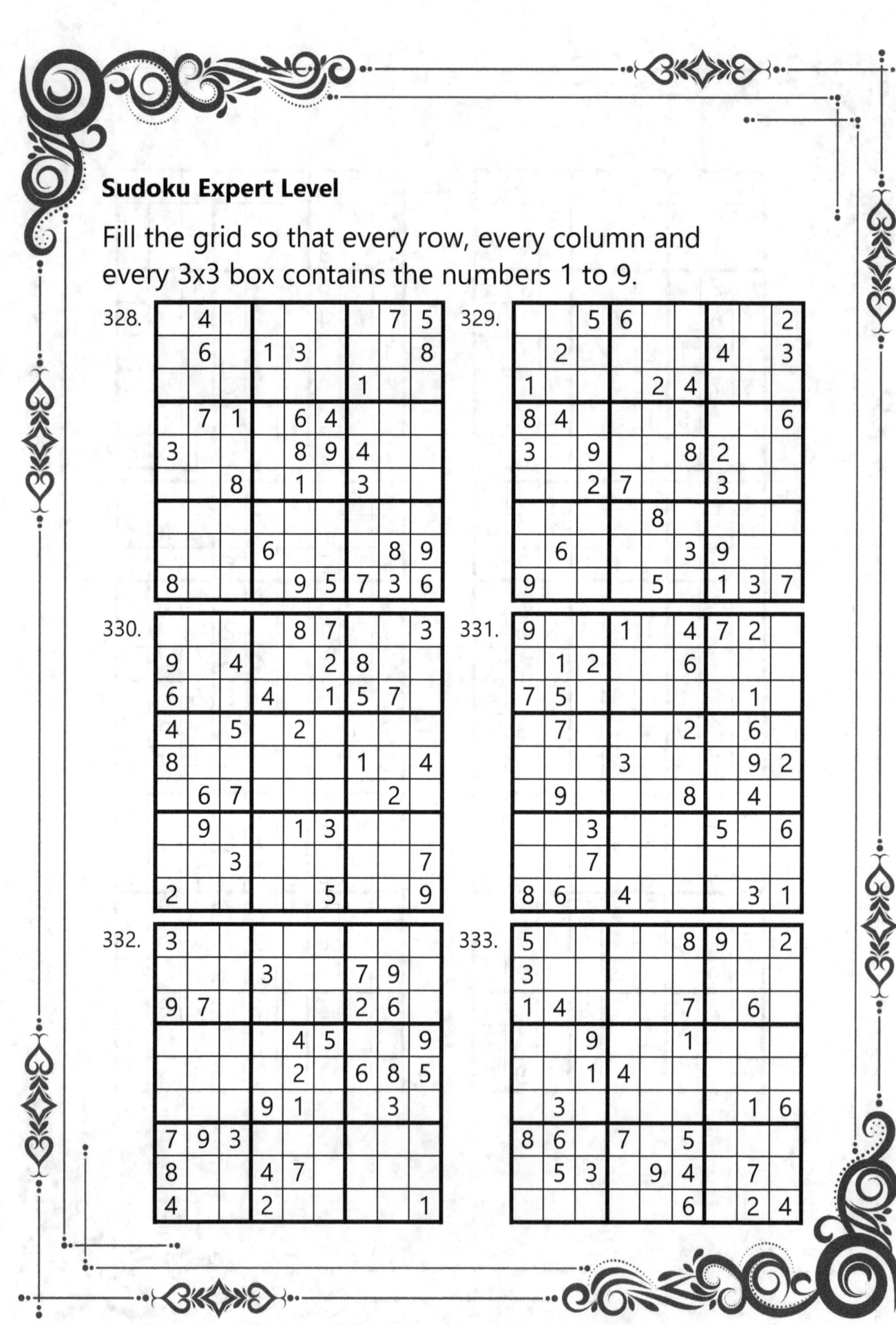

328.

	4					7	5	
	6		1	3				8
					1			
	7	1		6	4			
3				8	9	4		
		8		1		3		
			6			8	9	
8				9	5	7	3	6

329.

			5	6				2
	2					4		3
1				2	4			
8	4							6
3		9			8	2		
		2	7			3		
			8					
	6				3	9		
9				5		1	3	7

330.

				8	7			3
9		4			2	8		
6			4		1	5	7	
4		5		2				
8					1		4	
	6	7				2		
	9			1	3			
		3						7
2					5			9

331.

9			1		4	7	2	
	1	2			6			
7	5						1	
	7				2		6	
				3			9	2
	9				8		4	
			3			5		6
			7					
8	6		4				3	1

332.

3								
		3			7	9		
9	7				2	6		
			4	5				9
			2		6	8	5	
			9	1		3		
7	9	3						
8			4	7				
4			2					1

333.

5					8	9		2
3								
1	4				7		6	
		9			1			
		1	4					
	3						1	6
8	6		7		5			
	5	3		9	4		7	
					6		2	4

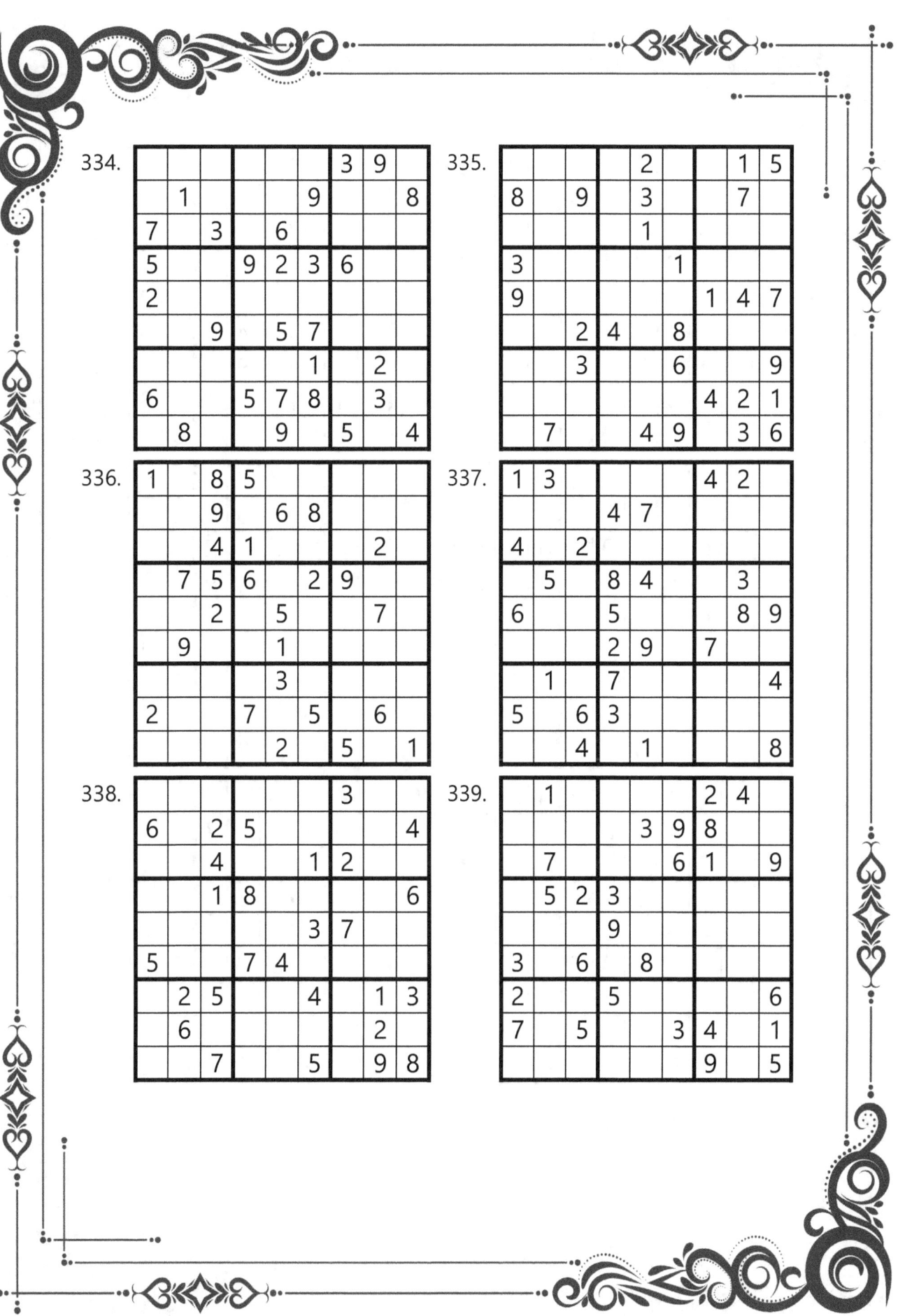

334.

					3	9		
	1			9				8
7		3		6				
5			9	2	3	6		
2								
		9		5	7			
					1		2	
6			5	7	8		3	
	8			9		5		4

335.

			2				1	5
8		9		3			7	
				1				
3					1			
9						1	4	7
		2	4		8			
		3			6			9
						4	2	1
	7			4	9		3	6

336.

1		8	5					
		9		6	8			
		4	1				2	
	7	5	6		2	9		
		2		5			7	
	9			1				
				3				
2			7		5		6	
				2		5		1

337.

1	3					4	2	
			4	7				
4		2						
	5		8	4			3	
6			5				8	9
			2	9		7		
	1		7					4
5		6	3					
		4		1				8

338.

						3		
6		2	5					4
		4			1	2		
		1	8					6
					3	7		
5			7	4				
	2	5			4		1	3
	6					2		
		7			5		9	8

339.

	1					2	4	
				3	9	8		
	7				6	1		9
	5	2	3					
			9					
3		6		8				
2			5					6
7		5			3	4		1
						9		5

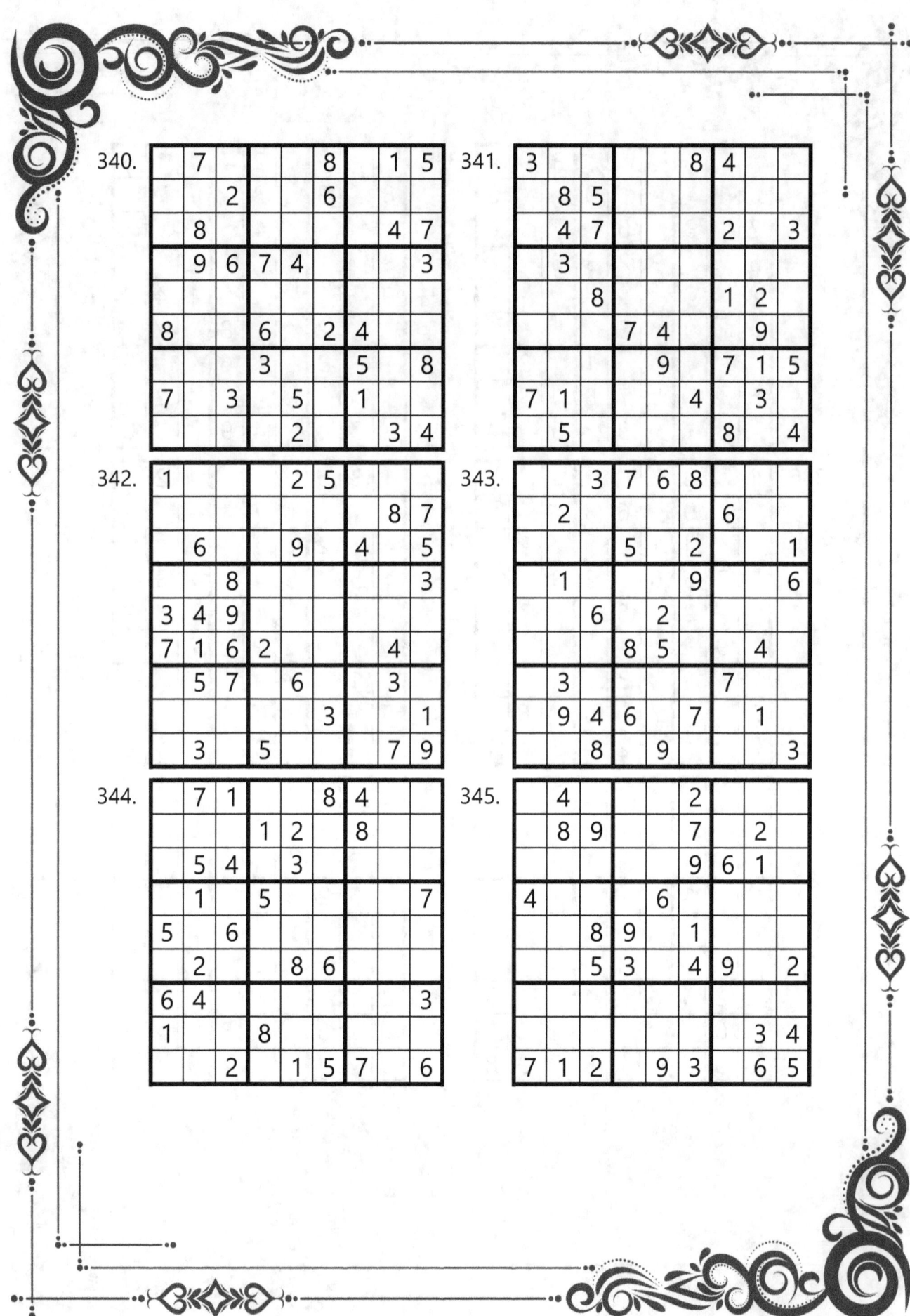

340.

7					8		1	5
	2				6			
	8						4	7
	9	6	7	4				3
8			6		2	4		
		3				5		8
7		3		5		1		
			2				3	4

341.

3					8	4		
	8	5						
	4	7				2		3
	3							
		8				1	2	
			7	4			9	
			9			7	1	5
7	1				4		3	
	5					8		4

342.

1				2	5			
							8	7
	6			9		4		5
		8						3
3	4	9						
7	1	6	2			4		
	5	7		6		3		
					3			1
	3		5			7	9	

343.

		3	7	6	8			
	2					6		
			5		2			1
	1				9			6
		6		2				
			8	5			4	
	3					7		
	9	4	6		7		1	
		8		9				3

344.

	7	1			8	4		
			1	2		8		
	5	4		3				
	1		5					7
5		6						
	2			8	6			
6	4							3
1			8					
		2		1	5	7		6

345.

	4				2			
	8	9			7		2	
					9	6	1	
4				6				
		8	9		1			
	5	3		4	9			2
							3	4
7	1	2		9	3		6	5

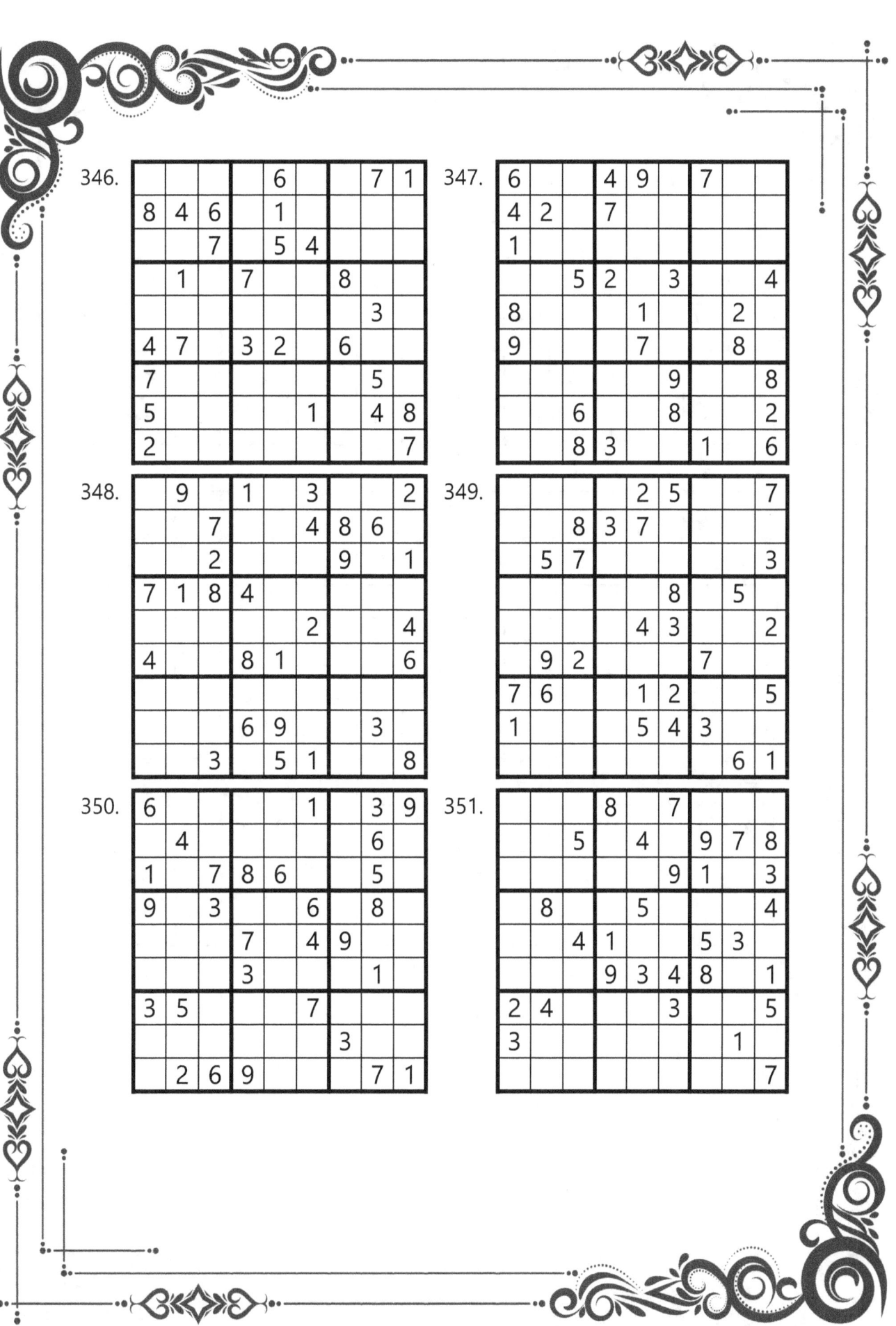

346.

				6			7	1
8	4	6		1				
		7		5	4			
	1		7			8		
						3		
4	7		3	2		6		
7						5		
5					1		4	8
2								7

347.

6			4	9		7		
4	2		7					
1								
	5	2		3				4
8			1			2		
9			7			8		
				9				8
	6			8				2
	8	3			1			6

348.

	9		1		3			2
		7		4	8	6		
		2			9		1	
7	1	8	4					
			2				4	
4			8	1			6	
		6	9			3		
		3		5	1		8	

349.

			2	5				7
		8	3	7				
	5	7						3
				8		5		
			4	3				2
	9	2				7		
7	6			1	2			5
1				5	4	3		
							6	1

350.

6					1		3	9
	4						6	
1			7	8	6		5	
9			3			6	8	
			7		4	9		
			3				1	
3	5			7				
						3		
		2	6	9			7	1

351.

			8		7			
		5		4		9	7	8
						9	1	3
	8			5				4
	4	1				5	3	
			9	3	4	8		1
2	4				3			5
3							1	
								7

Across-Downs Addition

Solve.

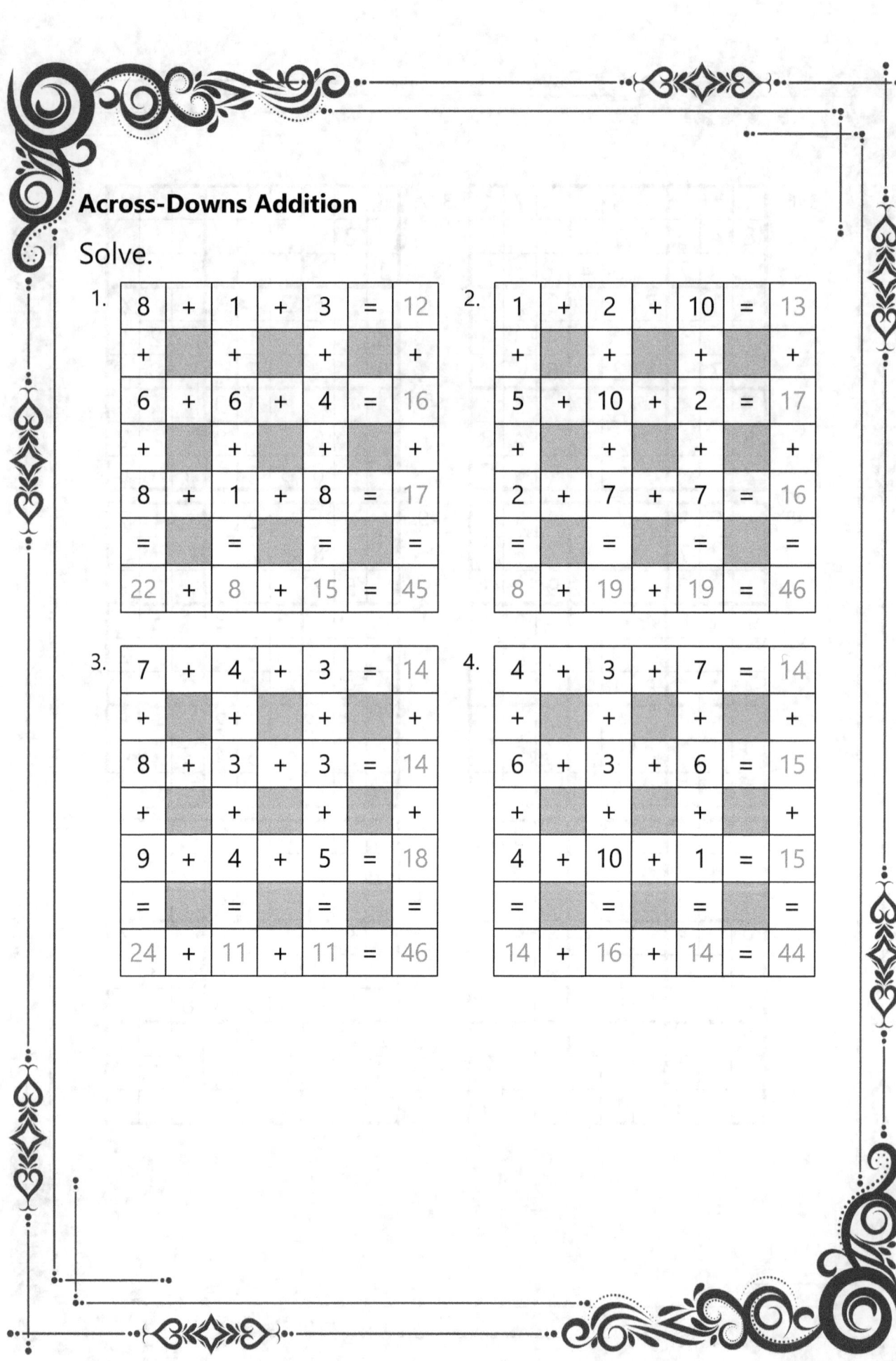

1.

8	+	1	+	3	=	12
+		+		+		+
6	+	6	+	4	=	16
+		+		+		+
8	+	1	+	8	=	17
=		=		=		=
22	+	8	+	15	=	45

2.

1	+	2	+	10	=	13
+		+		+		+
5	+	10	+	2	=	17
+		+		+		+
2	+	7	+	7	=	16
=		=		=		=
8	+	19	+	19	=	46

3.

7	+	4	+	3	=	14
+		+		+		+
8	+	3	+	3	=	14
+		+		+		+
9	+	4	+	5	=	18
=		=		=		=
24	+	11	+	11	=	46

4.

4	+	3	+	7	=	14
+		+		+		+
6	+	3	+	6	=	15
+		+		+		+
4	+	10	+	1	=	15
=		=		=		=
14	+	16	+	14	=	44

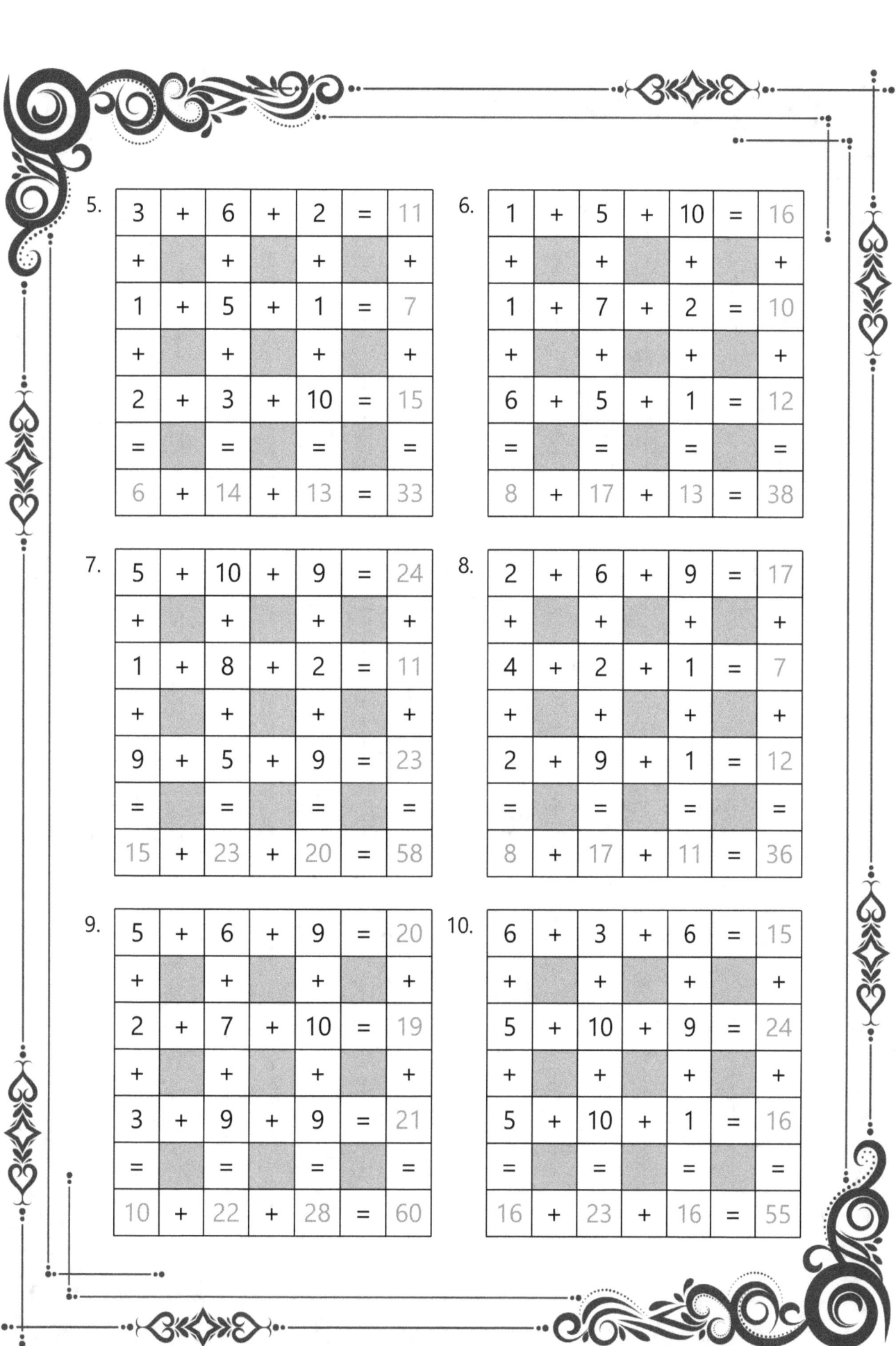

5.

3	+	6	+	2	=	11
+		+		+		+
1	+	5	+	1	=	7
+		+		+		+
2	+	3	+	10	=	15
=		=		=		=
6	+	14	+	13	=	33

6.

1	+	5	+	10	=	16
+		+		+		+
1	+	7	+	2	=	10
+		+		+		+
6	+	5	+	1	=	12
=		=		=		=
8	+	17	+	13	=	38

7.

5	+	10	+	9	=	24
+		+		+		+
1	+	8	+	2	=	11
+		+		+		+
9	+	5	+	9	=	23
=		=		=		=
15	+	23	+	20	=	58

8.

2	+	6	+	9	=	17
+		+		+		+
4	+	2	+	1	=	7
+		+		+		+
2	+	9	+	1	=	12
=		=		=		=
8	+	17	+	11	=	36

9.

5	+	6	+	9	=	20
+		+		+		+
2	+	7	+	10	=	19
+		+		+		+
3	+	9	+	9	=	21
=		=		=		=
10	+	22	+	28	=	60

10.

6	+	3	+	6	=	15
+		+		+		+
5	+	10	+	9	=	24
+		+		+		+
5	+	10	+	1	=	16
=		=		=		=
16	+	23	+	16	=	55

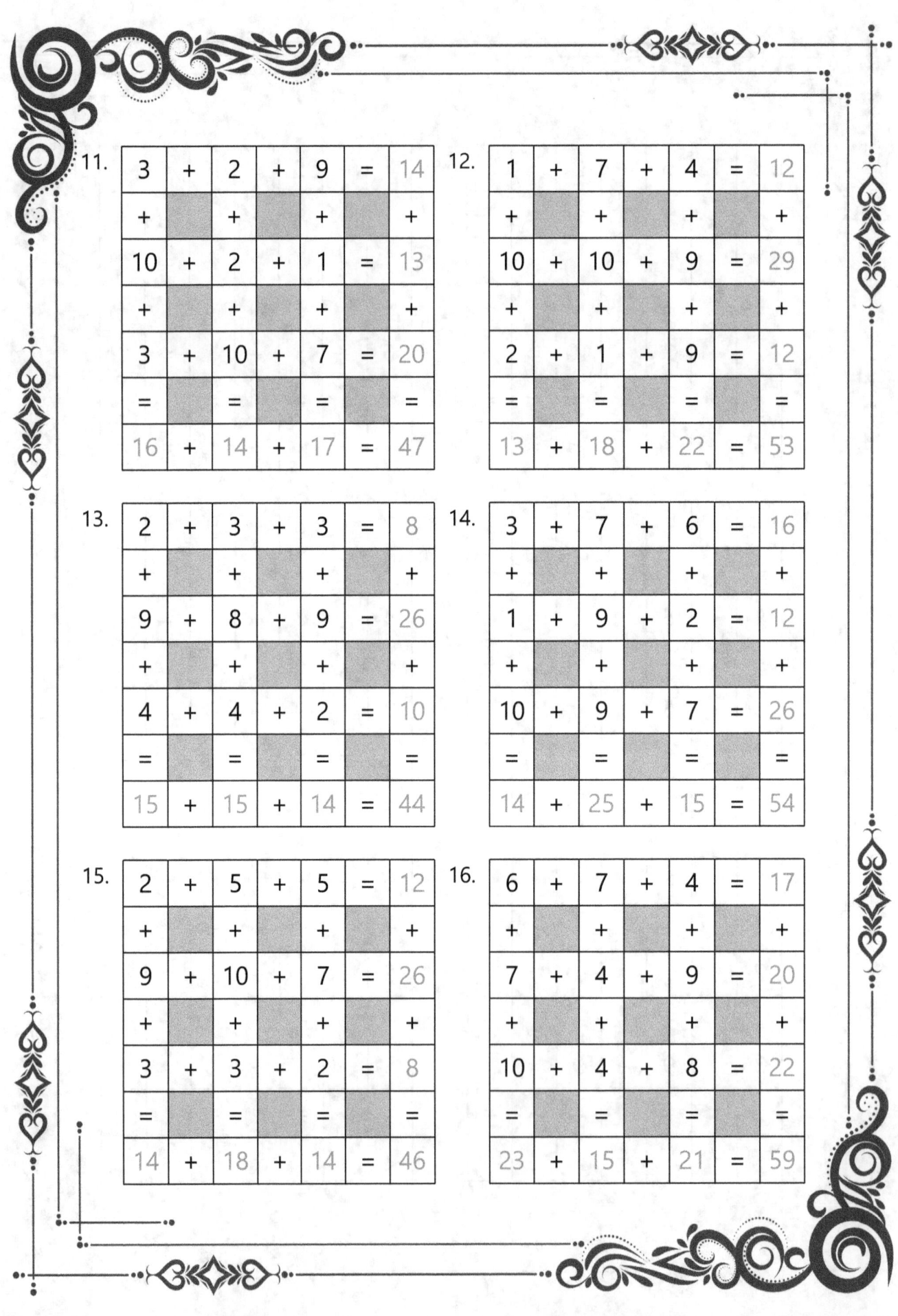

11.

3	+	2	+	9	=	14
+		+		+		+
10	+	2	+	1	=	13
+		+		+		+
3	+	10	+	7	=	20
=		=		=		=
16	+	14	+	17	=	47

12.

1	+	7	+	4	=	12
+		+		+		+
10	+	10	+	9	=	29
+		+		+		+
2	+	1	+	9	=	12
=		=		=		=
13	+	18	+	22	=	53

13.

2	+	3	+	3	=	8
+		+		+		+
9	+	8	+	9	=	26
+		+		+		+
4	+	4	+	2	=	10
=		=		=		=
15	+	15	+	14	=	44

14.

3	+	7	+	6	=	16
+		+		+		+
1	+	9	+	2	=	12
+		+		+		+
10	+	9	+	7	=	26
=		=		=		=
14	+	25	+	15	=	54

15.

2	+	5	+	5	=	12
+		+		+		+
9	+	10	+	7	=	26
+		+		+		+
3	+	3	+	2	=	8
=		=		=		=
14	+	18	+	14	=	46

16.

6	+	7	+	4	=	17
+		+		+		+
7	+	4	+	9	=	20
+		+		+		+
10	+	4	+	8	=	22
=		=		=		=
23	+	15	+	21	=	59

Across-Downs Subtraction

Solve.

17.

50	–	15	–	20	=	15
–		–		–		–
18	–	4	–	4	=	10
–		–		–		–
12	–	5	–	6	=	1
=		=		=		=
20	–	6	–	10	=	4

18.

61	–	25	–	20	=	16
–		–		–		–
18	–	8	–	7	=	3
–		–		–		–
21	–	9	–	6	=	6
=		=		=		=
22	–	8	–	7	=	7

19.

53	–	17	–	19	=	17
–		–		–		–
12	–	4	–	1	=	7
–		–		–		–
19	–	6	–	9	=	4
=		=		=		=
22	–	7	–	9	=	6

20.

46	–	15	–	13	=	18
–		–		–		–
18	–	10	–	1	=	7
–		–		–		–
16	–	2	–	9	=	5
=		=		=		=
12	–	3	–	3	=	6

21.

56	–	15	–	20	=	21
–		–		–		–
14	–	8	–	2	=	4
–		–		–		–
19	–	1	–	8	=	10
=		=		=		=
23	–	6	–	10	=	7

22.

33	–	7	–	19	=	7
–		–		–		–
13	–	4	–	6	=	3
–		–		–		–
15	–	2	–	10	=	3
=		=		=		=
5	–	1	–	3	=	1

23.

61	–	30	–	22	=	9
–		–		–		–
23	–	10	–	10	=	3
–		–		–		–
18	–	10	–	3	=	5
=		=		=		=
20	–	10	–	9	=	1

24.

50	–	18	–	13	=	19
–		–		–		–
21	–	9	–	5	=	7
–		–		–		–
10	–	1	–	5	=	4
=		=		=		=
19	–	8	–	3	=	8

25.

48	–	13	–	18	=	17
–		–		–		–
18	–	3	–	9	=	6
–		–		–		–
17	–	7	–	4	=	6
=		=		=		=
13	–	3	–	5	=	5

26.

40	–	16	–	13	=	11
–		–		–		–
21	–	7	–	8	=	6
–		–		–		–
9	–	2	–	3	=	4
=		=		=		=
10	–	7	–	2	=	1

27.

64	–	18	–	23	=	23
–		–		–		–
23	–	7	–	9	=	7
–		–		–		–
27	–	8	–	10	=	9
=		=		=		=
14	–	3	–	4	=	7

28.

64	–	20	–	24	=	20
–		–		–		–
25	–	7	–	9	=	9
–		–		–		–
21	–	6	–	6	=	9
=		=		=		=
18	–	7	–	9	=	2

29.

60	–	25	–	18	=	17
–		–		–		–
16	–	7	–	7	=	2
–		–		–		–
23	–	10	–	6	=	7
=		=		=		=
21	–	8	–	5	=	8

30.

46	–	13	–	20	=	13
–		–		–		–
8	–	3	–	3	=	2
–		–		–		–
15	–	1	–	7	=	7
=		=		=		=
23	–	9	–	10	=	4

31.

40	–	13	–	16	=	11
–		–		–		–
21	–	9	–	6	=	6
–		–		–		–
6	–	2	–	2	=	2
=		=		=		=
13	–	2	–	8	=	3

32.

51	–	16	–	19	=	16
–		–		–		–
22	–	6	–	7	=	9
–		–		–		–
17	–	9	–	5	=	3
=		=		=		=
12	–	1	–	7	=	4

Across-Downs Mixte

Solve.

33.

7	−	6	+	9	=	10
−		+		−		+
6	+	8	−	7	=	7
+		−		+		+
7	−	1	+	9	=	15
=		=		=		=
8	+	13	+	11	=	32

34.

7	−	3	+	1	=	5
−		+		−		+
3	+	5	−	1	=	7
+		−		+		+
3	−	1	+	4	=	6
=		=		=		=
7	+	7	+	4	=	18

35.

7	−	3	+	4	=	8
−		+		−		+
3	+	5	−	2	=	6
+		−		+		+
2	−	1	+	4	=	5
=		=		=		=
6	+	7	+	6	=	19

36.

10	−	7	+	5	=	8
−		+		−		+
7	+	8	−	5	=	10
+		−		+		+
4	−	1	+	2	=	5
=		=		=		=
7	+	14	+	2	=	23

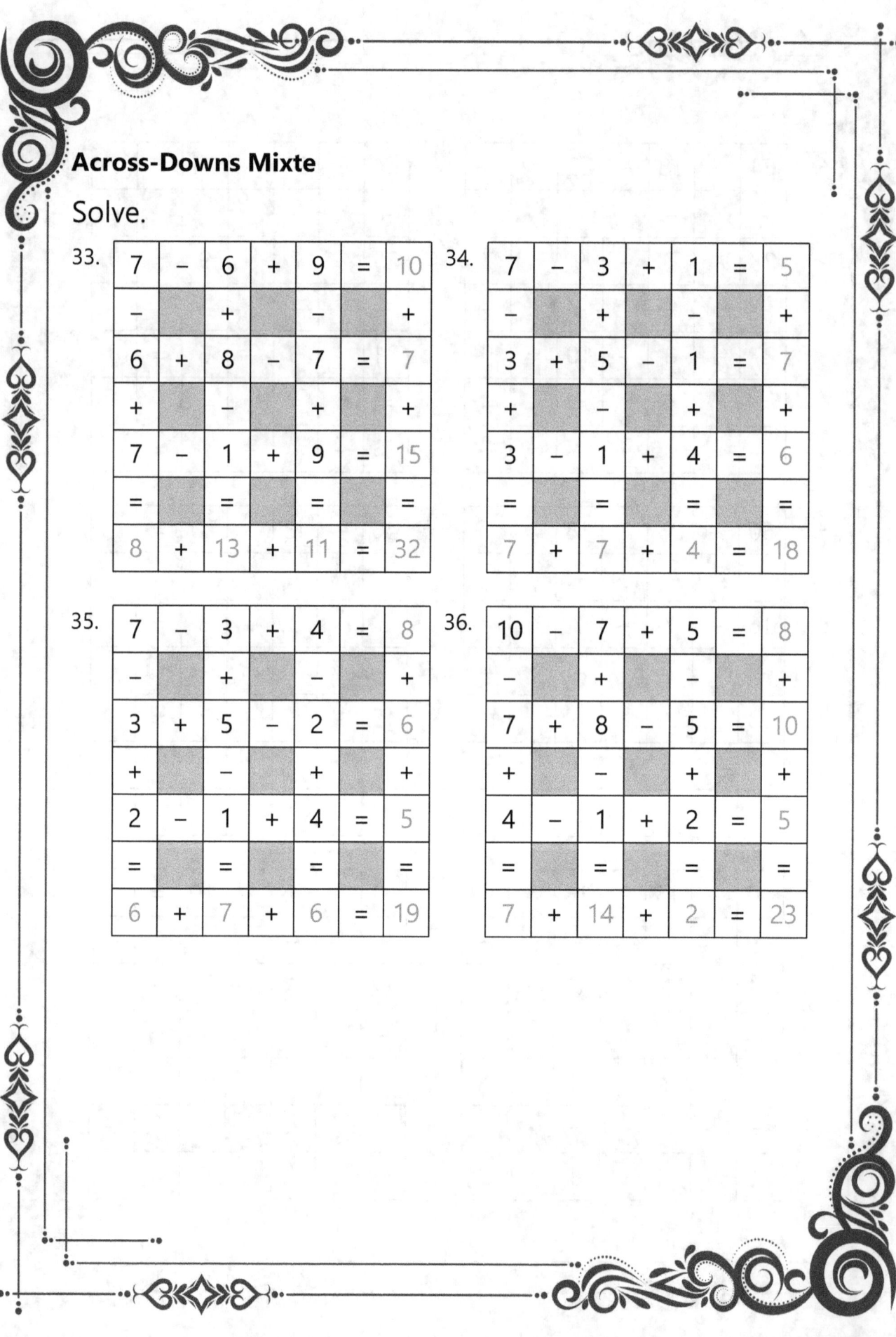

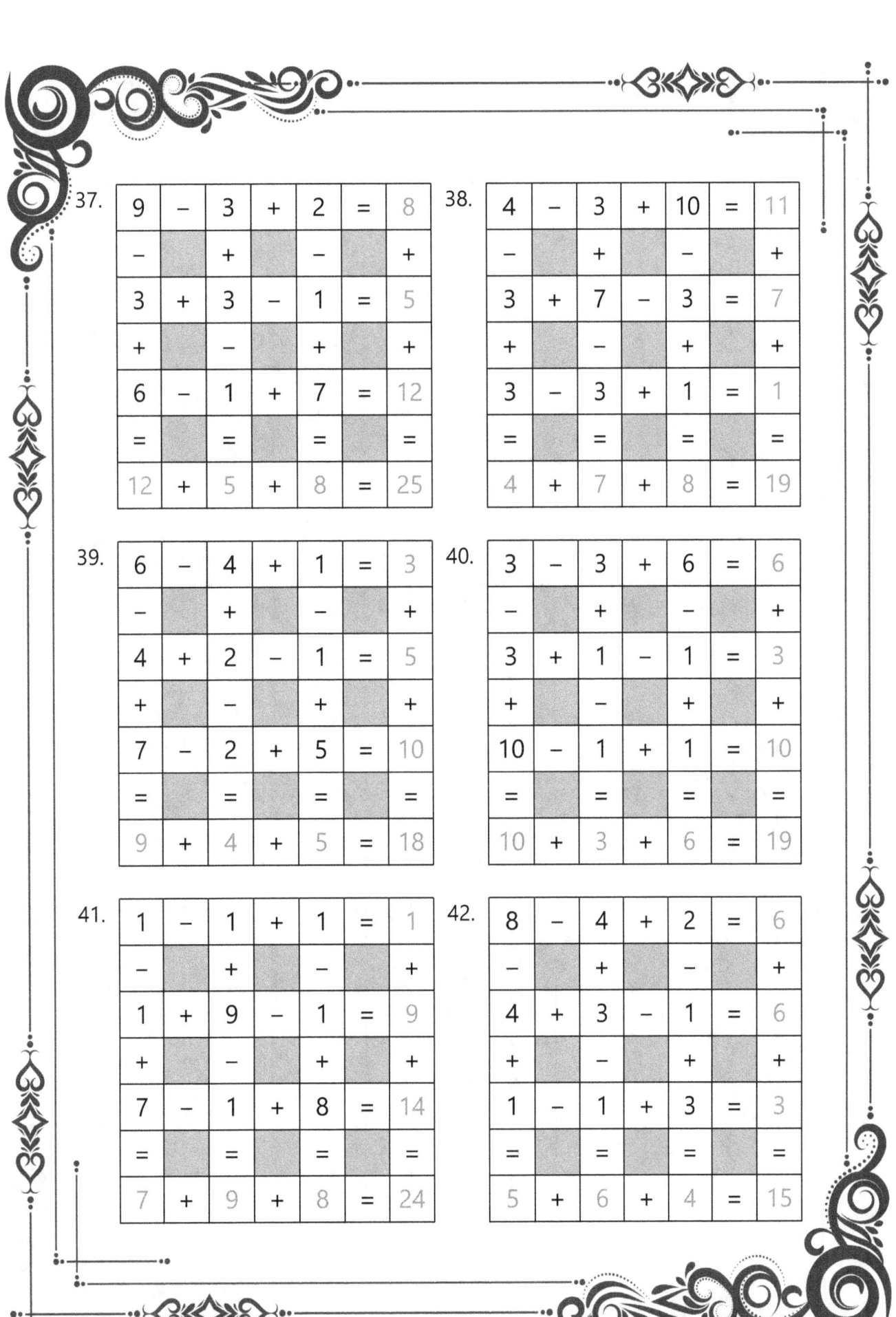

37.

9	–	3	+	2	=	8
–		+		–		+
3	+	3	–	1	=	5
+		–		+		+
6	–	1	+	7	=	12
=		=		=		=
12	+	5	+	8	=	25

38.

4	–	3	+	10	=	11
–		+		–		+
3	+	7	–	3	=	7
+		–		+		+
3	–	3	+	1	=	1
=		=		=		=
4	+	7	+	8	=	19

39.

6	–	4	+	1	=	3
–		+		–		+
4	+	2	–	1	=	5
+		–		+		+
7	–	2	+	5	=	10
=		=		=		=
9	+	4	+	5	=	18

40.

3	–	3	+	6	=	6
–		+		–		+
3	+	1	–	1	=	3
+		–		+		+
10	–	1	+	1	=	10
=		=		=		=
10	+	3	+	6	=	19

41.

1	–	1	+	1	=	1
–		+		–		+
1	+	9	–	1	=	9
+		–		+		+
7	–	1	+	8	=	14
=		=		=		=
7	+	9	+	8	=	24

42.

8	–	4	+	2	=	6
–		+		–		+
4	+	3	–	1	=	6
+		–		+		+
1	–	1	+	3	=	3
=		=		=		=
5	+	6	+	4	=	15

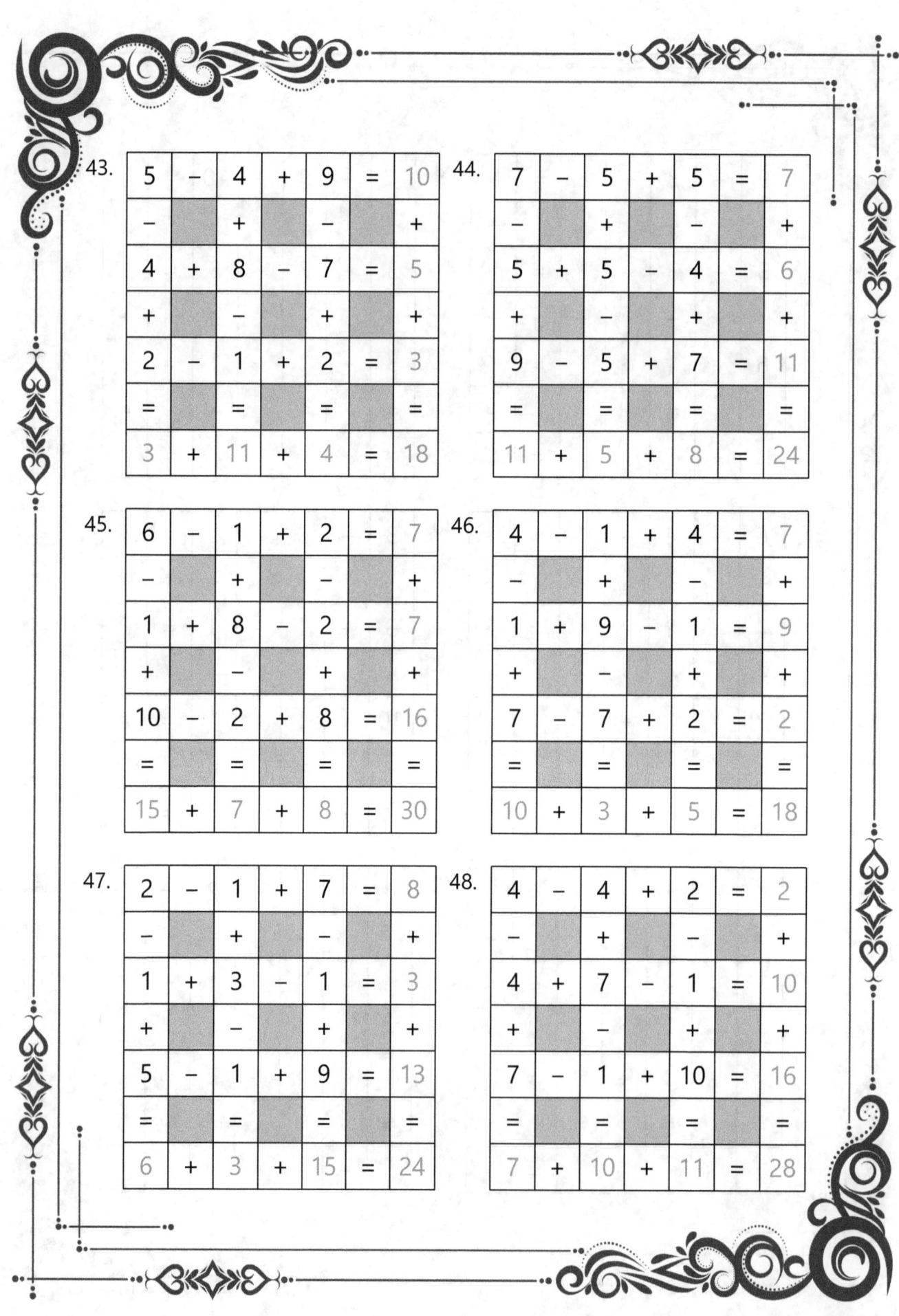

43.

5	–	4	+	9	=	10
–		+		–		+
4	+	8	–	7	=	5
+		–		+		+
2	–	1	+	2	=	3
=		=		=		=
3	+	11	+	4	=	18

44.

7	–	5	+	5	=	7
–		+		–		+
5	+	5	–	4	=	6
+		–		+		+
9	–	5	+	7	=	11
=		=		=		=
11	+	5	+	8	=	24

45.

6	–	1	+	2	=	7
–		+		–		+
1	+	8	–	2	=	7
+		–		+		+
10	–	2	+	8	=	16
=		=		=		=
15	+	7	+	8	=	30

46.

4	–	1	+	4	=	7
–		+		–		+
1	+	9	–	1	=	9
+		–		+		+
7	–	7	+	2	=	2
=		=		=		=
10	+	3	+	5	=	18

47.

2	–	1	+	7	=	8
–		+		–		+
1	+	3	–	1	=	3
+		–		+		+
5	–	1	+	9	=	13
=		=		=		=
6	+	3	+	15	=	24

48.

4	–	4	+	2	=	2
–		+		–		+
4	+	7	–	1	=	10
+		–		+		+
7	–	1	+	10	=	16
=		=		=		=
7	+	10	+	11	=	28

49.

5	−	3	+	8	=	10
−		+		−		+
3	+	4	−	4	=	3
+		−		+		+
10	−	3	+	5	=	12
=		=		=		=
12	+	4	+	9	=	25

50.

7	−	2	+	4	=	9
−		+		−		+
2	+	8	−	4	=	6
+		−		+		+
6	−	1	+	1	=	6
=		=		=		=
11	+	9	+	1	=	21

51.

7	−	3	+	3	=	7
−		+		−		+
3	+	5	−	1	=	7
+		−		+		+
7	−	3	+	2	=	6
=		=		=		=
11	+	5	+	4	=	20

52.

8	−	3	+	4	=	9
−		+		−		+
3	+	7	−	4	=	6
+		−		+		+
5	−	3	+	8	=	10
=		=		=		=
10	+	7	+	8	=	25

53.

1	−	1	+	4	=	4
−		+		−		+
1	+	6	−	4	=	3
+		−		+		+
7	−	2	+	3	=	8
=		=		=		=
7	+	5	+	3	=	15

54.

8	−	2	+	7	=	13
−		+		−		+
2	+	9	−	1	=	10
+		−		+		+
6	−	1	+	8	=	13
=		=		=		=
12	+	10	+	14	=	36

55.

10	–	2	+	1	=	9
–		+		–		+
2	+	9	–	1	=	10
+		–		+		+
7	–	5	+	8	=	10
=		=		=		=
15	+	6	+	8	=	29

56.

10	–	7	+	6	=	9
–		+		–		+
7	+	7	–	4	=	10
+		–		+		+
10	–	4	+	2	=	8
=		=		=		=
13	+	10	+	4	=	27

57.

1	–	1	+	8	=	8
–		+		–		+
1	+	1	–	1	=	1
+		–		+		+
7	–	1	+	2	=	8
=		=		=		=
7	+	1	+	9	=	17

58.

3	–	2	+	8	=	9
–		+		–		+
2	+	8	–	4	=	6
+		–		+		+
7	–	3	+	4	=	8
=		=		=		=
8	+	7	+	8	=	23

59.

8	–	6	+	6	=	8
–		+		–		+
6	+	4	–	1	=	9
+		–		+		+
6	–	3	+	9	=	12
=		=		=		=
8	+	7	+	14	=	29

60.

10	–	1	+	8	=	17
–		+		–		+
1	+	4	–	4	=	1
+		–		+		+
7	–	3	+	3	=	7
=		=		=		=
16	+	2	+	7	=	25

Magic Squares

Find the magic number.

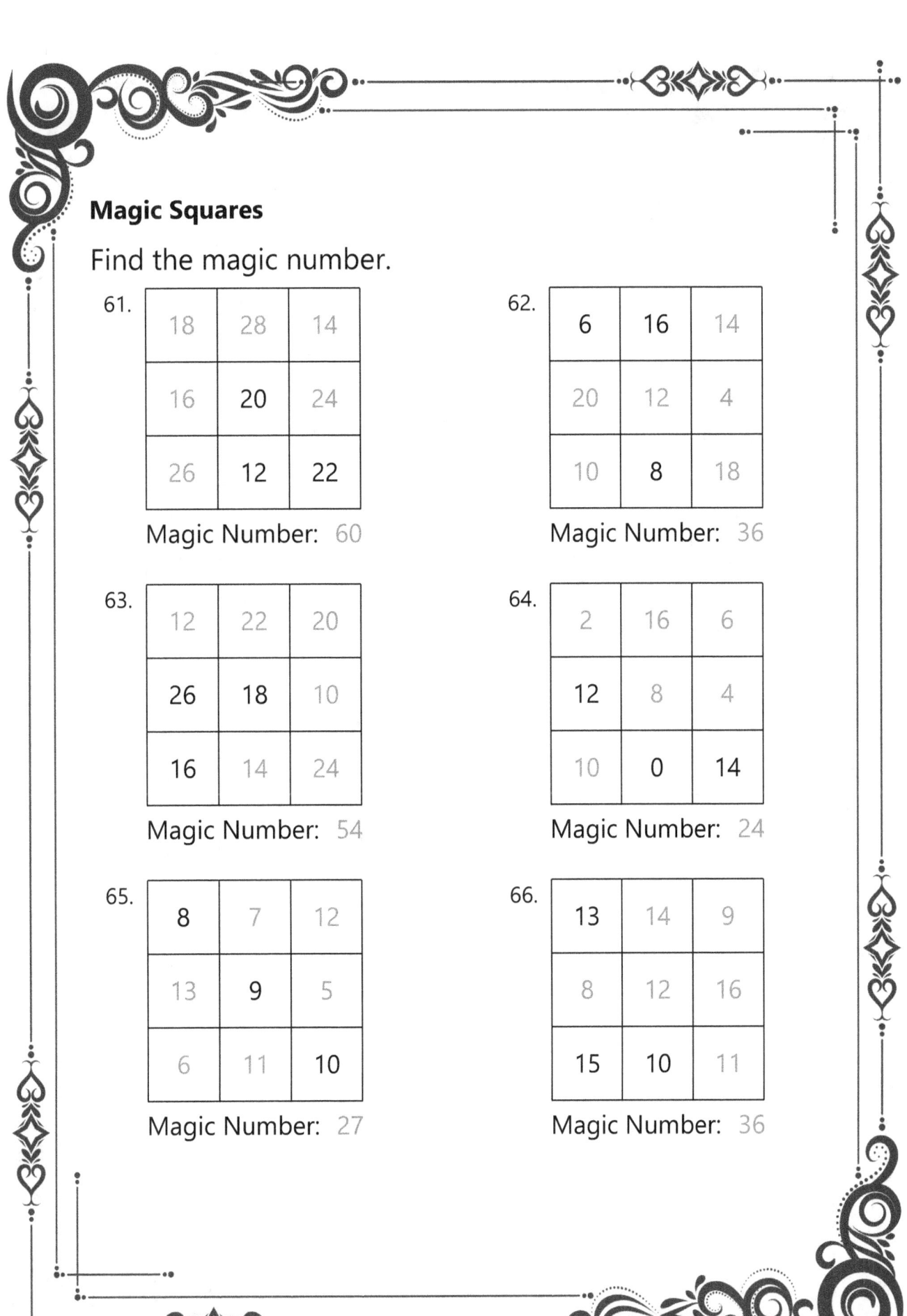

61.

18	28	14
16	20	24
26	12	22

Magic Number: 60

62.

6	16	14
20	12	4
10	8	18

Magic Number: 36

63.

12	22	20
26	18	10
16	14	24

Magic Number: 54

64.

2	16	6
12	8	4
10	0	14

Magic Number: 24

65.

8	7	12
13	9	5
6	11	10

Magic Number: 27

66.

13	14	9
8	12	16
15	10	11

Magic Number: 36

67.

12	**11**	16
17	13	9
10	15	**14**

Magic Number: 39

68.

20	**34**	24
30	**26**	22
28	18	32

Magic Number: 78

69.

10	**12**	2
0	**8**	16
14	4	6

Magic Number: 24

70.

16	11	**12**
9	**13**	17
14	15	10

Magic Number: 39

71.

3	8	**7**
10	**6**	**2**
5	4	9

Magic Number: 18

72.

6	11	4
5	7	**9**
10	**3**	8

Magic Number: 21

73.

1	8	3
6	4	2
5	0	7

Magic Number: 12

74.

28	18	32
30	26	22
20	34	24

Magic Number: 78

75.

8	3	4
1	5	9
6	7	2

Magic Number: 15

76.

16	26	12
14	18	22
24	10	20

Magic Number: 54

77.

5	10	9
12	8	4
7	6	11

Magic Number: 24

78.

2	7	6
9	5	1
4	3	8

Magic Number: 15

79.

28	14	24
18	22	26
20	30	16

Magic Number: 66

80.

12	7	14
13	11	9
8	15	10

Magic Number: 33

81.

20	10	24
22	18	14
12	26	16

Magic Number: 54

82.

12	2	16
14	10	6
4	18	8

Magic Number: 30

83.

9	4	11
10	8	6
5	12	7

Magic Number: 24

84.

10	24	14
20	16	12
18	8	22

Magic Number: 48

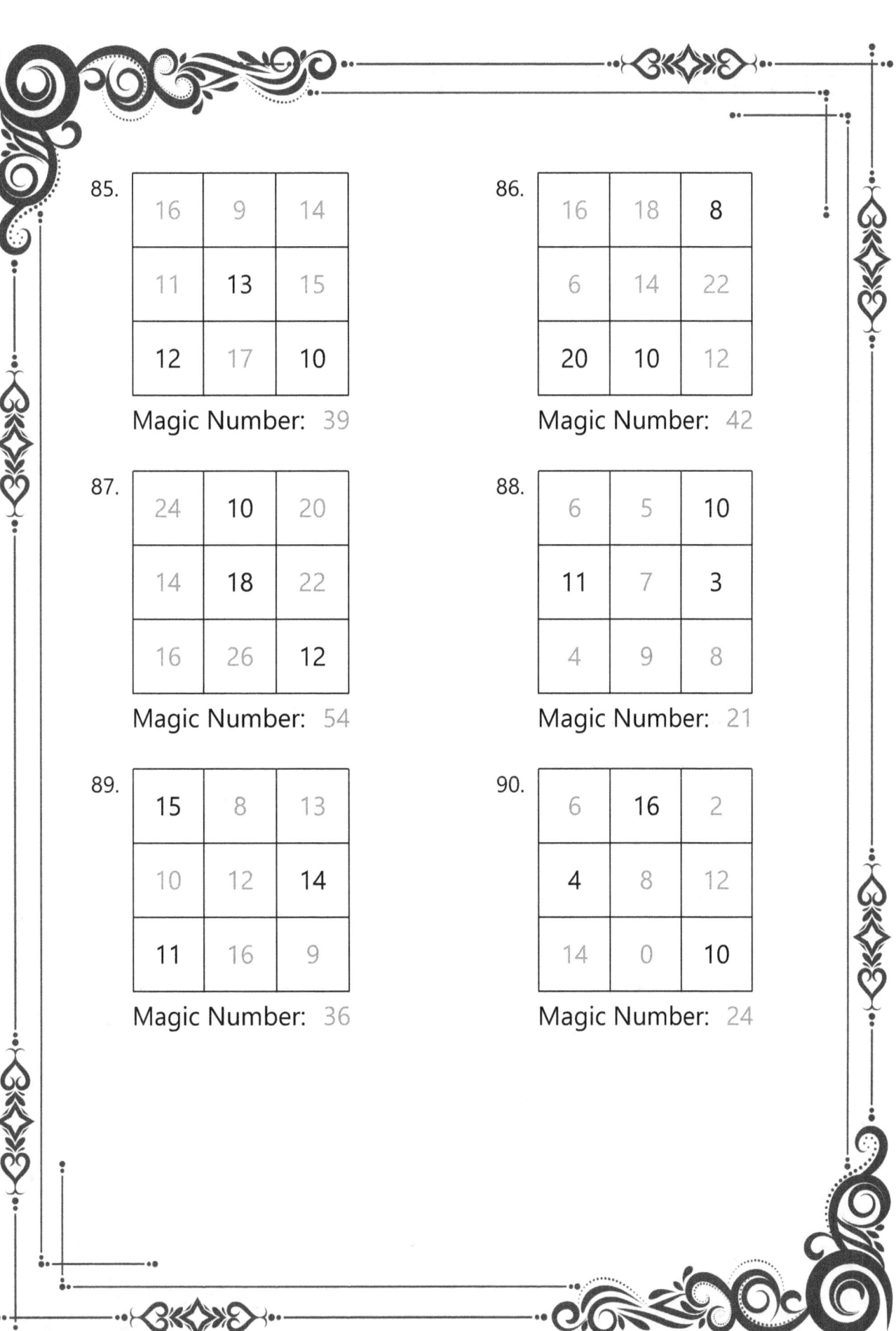

85.

16	9	14
11	**13**	15
12	17	**10**

Magic Number: 39

86.

16	18	**8**
6	14	22
20	**10**	12

Magic Number: 42

87.

24	**10**	20
14	**18**	22
16	26	**12**

Magic Number: 54

88.

6	5	**10**
11	7	**3**
4	9	8

Magic Number: 21

89.

15	8	13
10	12	**14**
11	16	9

Magic Number: 36

90.

6	**16**	2
4	8	12
14	0	**10**

Magic Number: 24

91.

4	9	8
11	7	3
6	5	10

Magic Number: 21

92.

9	2	7
4	6	8
5	10	3

Magic Number: 18

93.

24	34	20
22	26	30
32	18	28

Magic Number: 78

94.

11	6	13
12	10	8
7	14	9

Magic Number: 30

95.

26	28	18
16	24	32
30	20	22

Magic Number: 72

96.

5	4	9
10	6	2
3	8	7

Magic Number: 18

97.

4	11	6
9	7	5
8	3	10

Magic Number: 21

98.

8	15	10
13	11	9
12	7	14

Magic Number: 33

99.

10	11	6
5	9	13
12	7	8

Magic Number: 27

100.

5	6	1
0	4	8
7	2	3

Magic Number: 12

101.

12	26	16
22	18	14
20	10	24

Magic Number: 54

102.

3	2	7
8	4	0
1	6	5

Magic Number: 12

103.

14	4	18
16	12	8
6	20	10

Magic Number: 36

104.

10	5	12
11	9	7
6	13	8

Magic Number: 27

105.

10	20	18
24	16	8
14	12	22

Magic Number: 48

106.

22	20	30
32	24	16
18	28	26

Magic Number: 72

107.

14	9	10
7	11	15
12	13	8

Magic Number: 33

108.

12	17	10
11	13	15
16	9	14

Magic Number: 39

109.

6	20	10
16	12	8
14	4	18

Magic Number: 36

110.

26	12	22
16	20	24
18	28	14

Magic Number: 60

111.

18	32	22
28	24	20
26	16	30

Magic Number: 72

112.

10	20	6
8	12	16
18	4	14

Magic Number: 36

113.

7	6	11
12	8	4
5	10	9

Magic Number: 24

114.

11	6	7
4	8	12
9	10	5

Magic Number: 24

115.

10	**8**	18
20	**12**	4
6	16	**14**

Magic Number: 36

116.

12	**13**	8
7	11	**15**
14	9	**10**

Magic Number: 33

117.

14	**16**	**6**
4	12	**20**
18	8	10

Magic Number: 36

118.

11	10	15
16	12	**8**
9	14	13

Magic Number: 36

119.

16	26	24
30	22	**14**
20	**18**	**28**

Magic Number: 66

120.

14	7	12
9	**11**	13
10	**15**	8

Magic Number: 33

Multiplication Box

Solve.

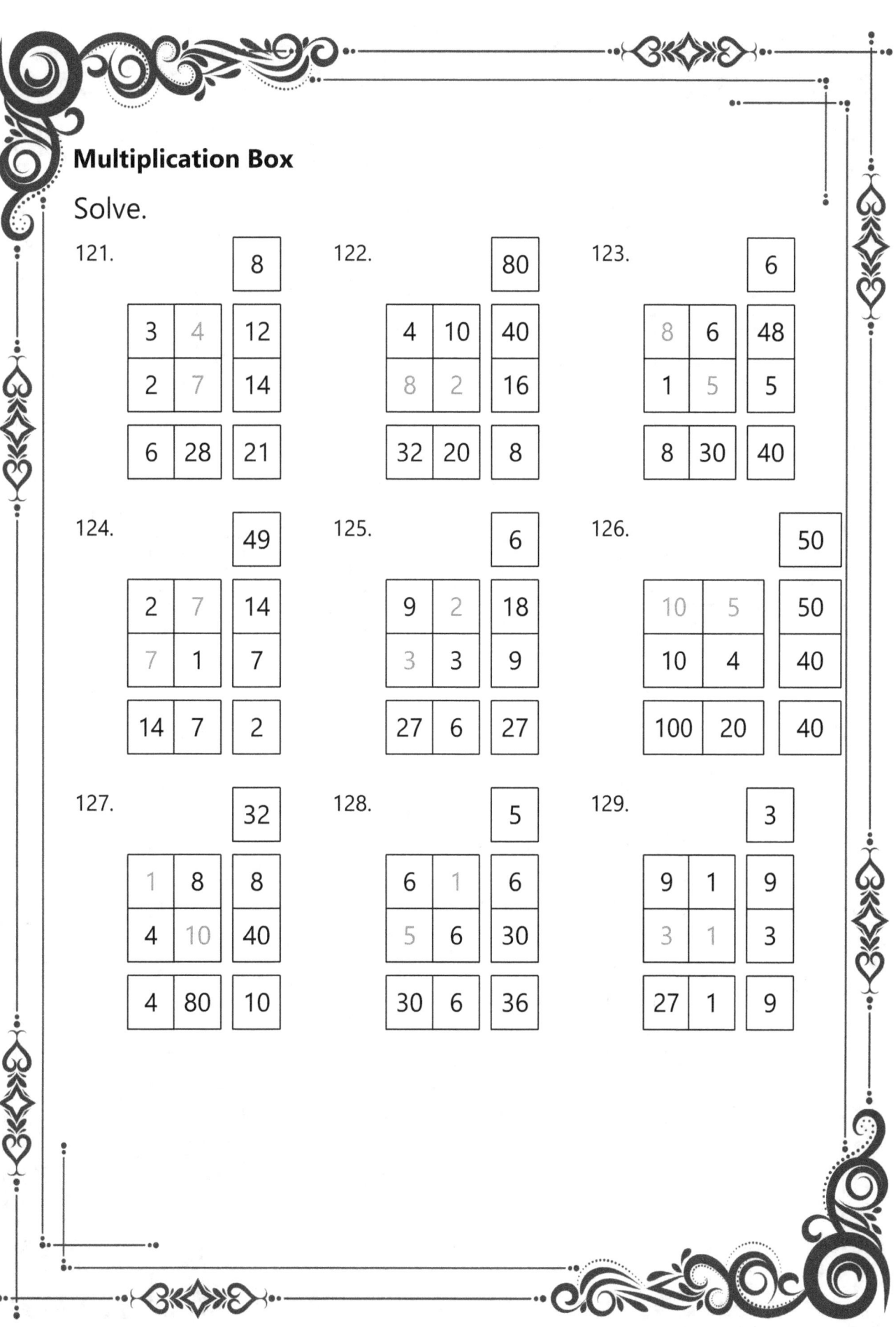

121.

		8
3	4	12
2	7	14
6	28	21

122.

		80
4	10	40
8	2	16
32	20	8

123.

		6
8	6	48
1	5	5
8	30	40

124.

		49
2	7	14
7	1	7
14	7	2

125.

		6
9	2	18
3	3	9
27	6	27

126.

		50
10	5	50
10	4	40
100	20	40

127.

		32
1	8	8
4	10	40
4	80	10

128.

		5
6	1	6
5	6	30
30	6	36

129.

		3
9	1	9
3	1	3
27	1	9

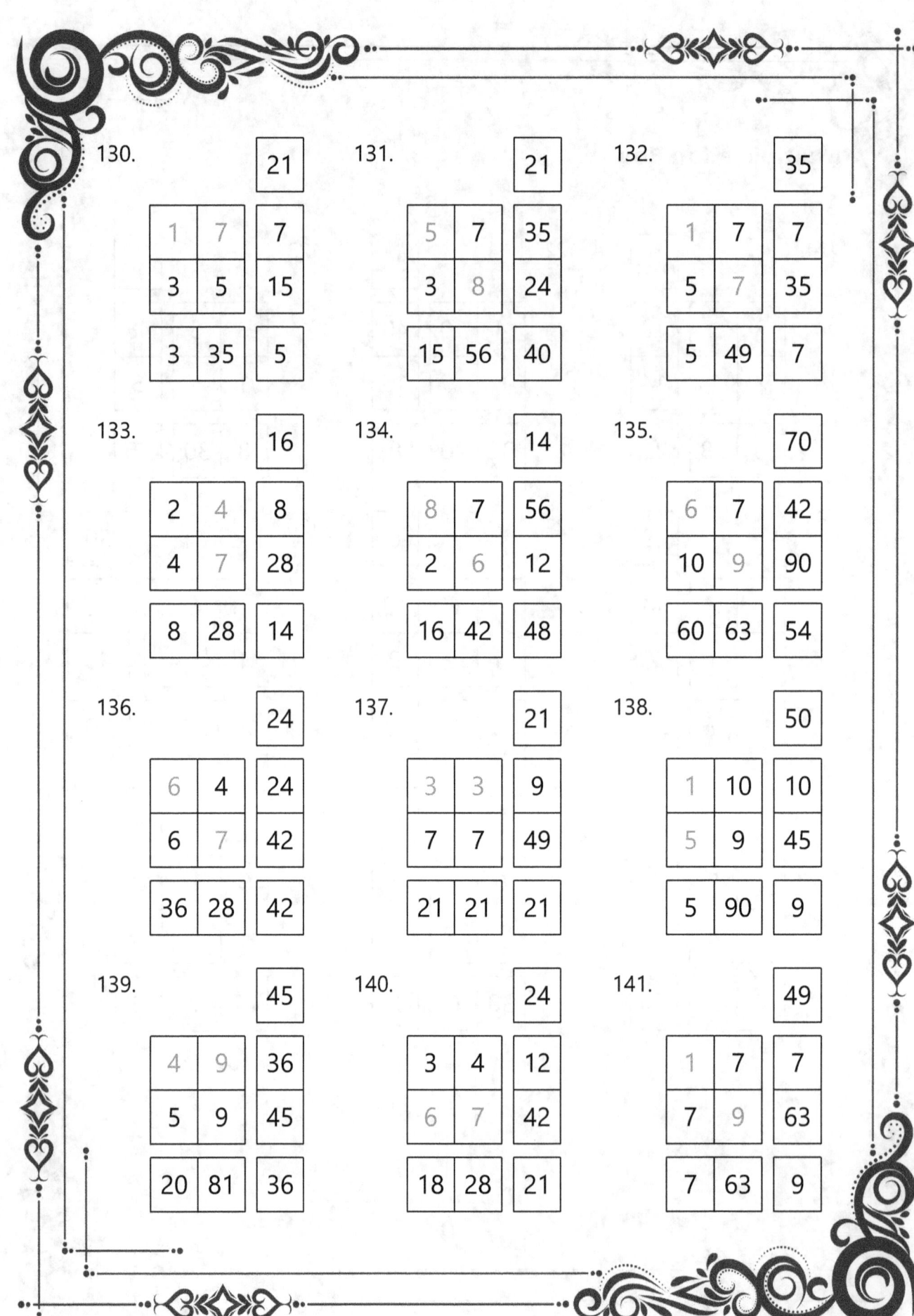

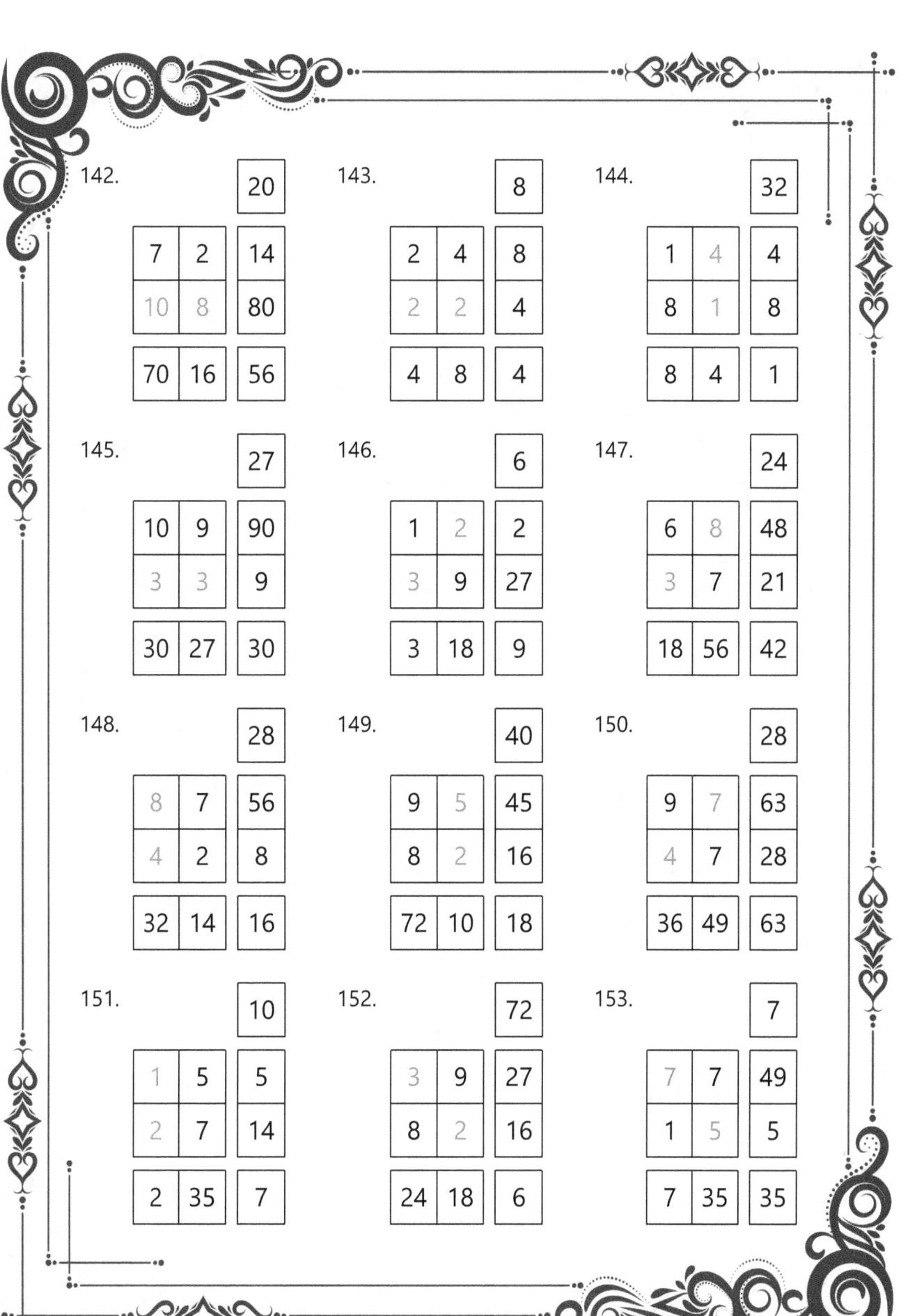

142.

		20
7	2	14
10	8	80
70	16	56

143.

		8
2	4	8
2	2	4
4	8	4

144.

		32
1	4	4
8	1	8
8	4	1

145.

		27
10	9	90
3	3	9
30	27	30

146.

		6
1	2	2
3	9	27
3	18	9

147.

		24
6	8	48
3	7	21
18	56	42

148.

		28
8	7	56
4	2	8
32	14	16

149.

		40
9	5	45
8	2	16
72	10	18

150.

		28
9	7	63
4	7	28
36	49	63

151.

		10
1	5	5
2	7	14
2	35	7

152.

		72
3	9	27
8	2	16
24	18	6

153.

		7
7	7	49
1	5	5
7	35	35

Secret Trails Addition

Find the secret trail.

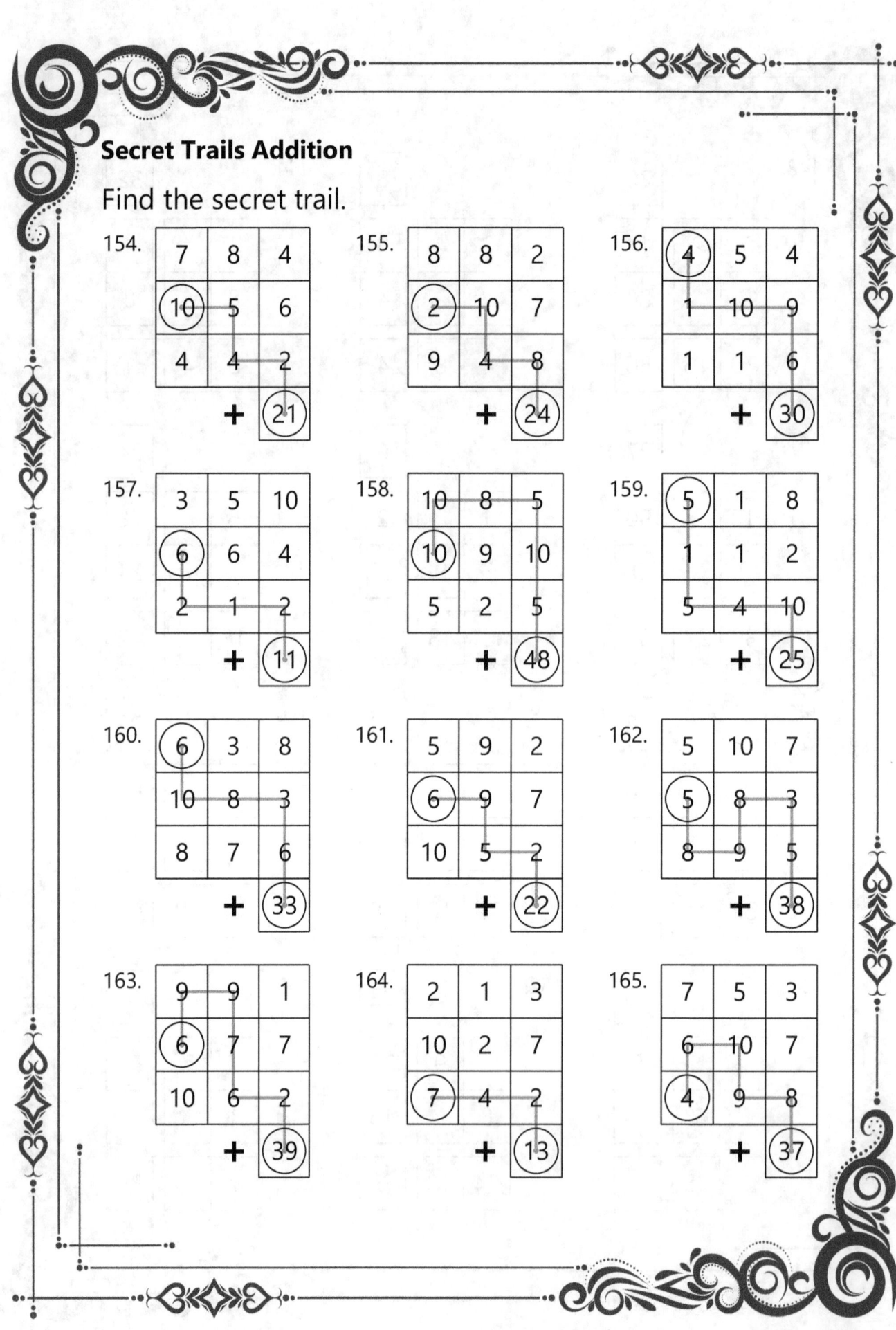

154.

7	8	4
10	5	6
4	4	2
	+	21

155.

8	8	2
2	10	7
9	4	8
	+	24

156.

4	5	4
1	10	9
1	1	6
	+	30

157.

3	5	10
6	6	4
2	1	2
	+	11

158.

10	8	5
10	9	10
5	2	5
	+	48

159.

5	1	8
1	1	2
5	4	10
	+	25

160.

6	3	8
10	8	3
8	7	6
	+	33

161.

5	9	2
6	9	7
10	5	2
	+	22

162.

5	10	7
5	8	3
8	9	5
	+	38

163.

9	9	1
6	7	7
10	6	2
	+	39

164.

2	1	3
10	2	7
7	4	2
	+	13

165.

7	5	3
6	10	7
4	9	8
	+	37

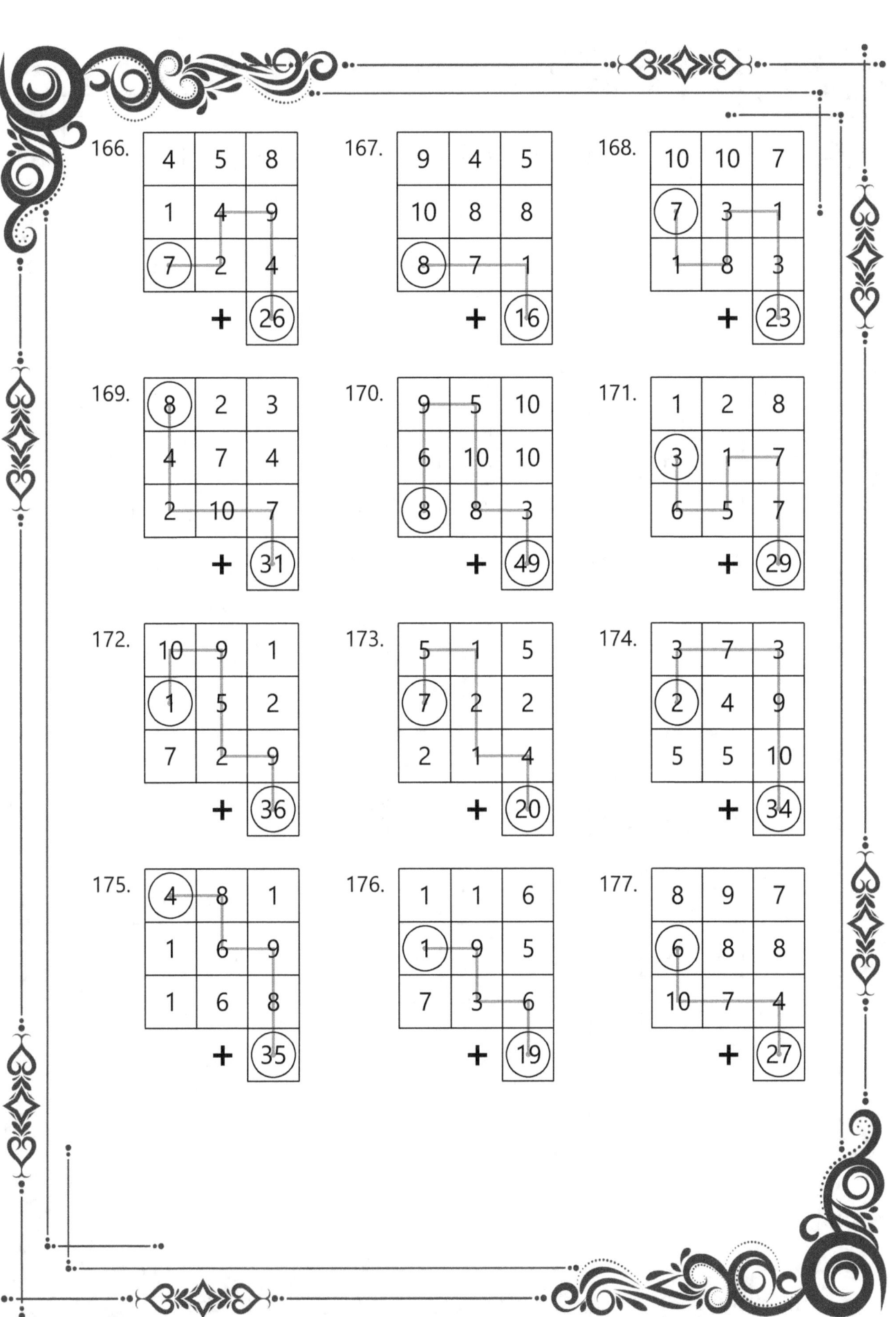

166.

4	5	8
1	4	9
(7)	2	4

+ (26)

167.

9	4	5
10	8	8
(8)	7	1

+ (16)

168.

10	10	7
(7)	3	1
1	8	3

+ (23)

169.

(8)	2	3
4	7	4
2	10	7

+ (31)

170.

9	5	10
6	10	10
(8)	8	3

+ (49)

171.

1	2	8
(3)	1	7
6	5	7

+ (29)

172.

10	9	1
(1)	5	2
7	2	9

+ (36)

173.

5	1	5
(7)	2	2
2	1	4

+ (20)

174.

3	7	3
(2)	4	9
5	5	10

+ (34)

175.

(4)	8	1
1	6	9
1	6	8

+ (35)

176.

1	1	6
(1)	9	5
7	3	6

+ (19)

177.

8	9	7
(6)	8	8
10	7	4

+ (27)

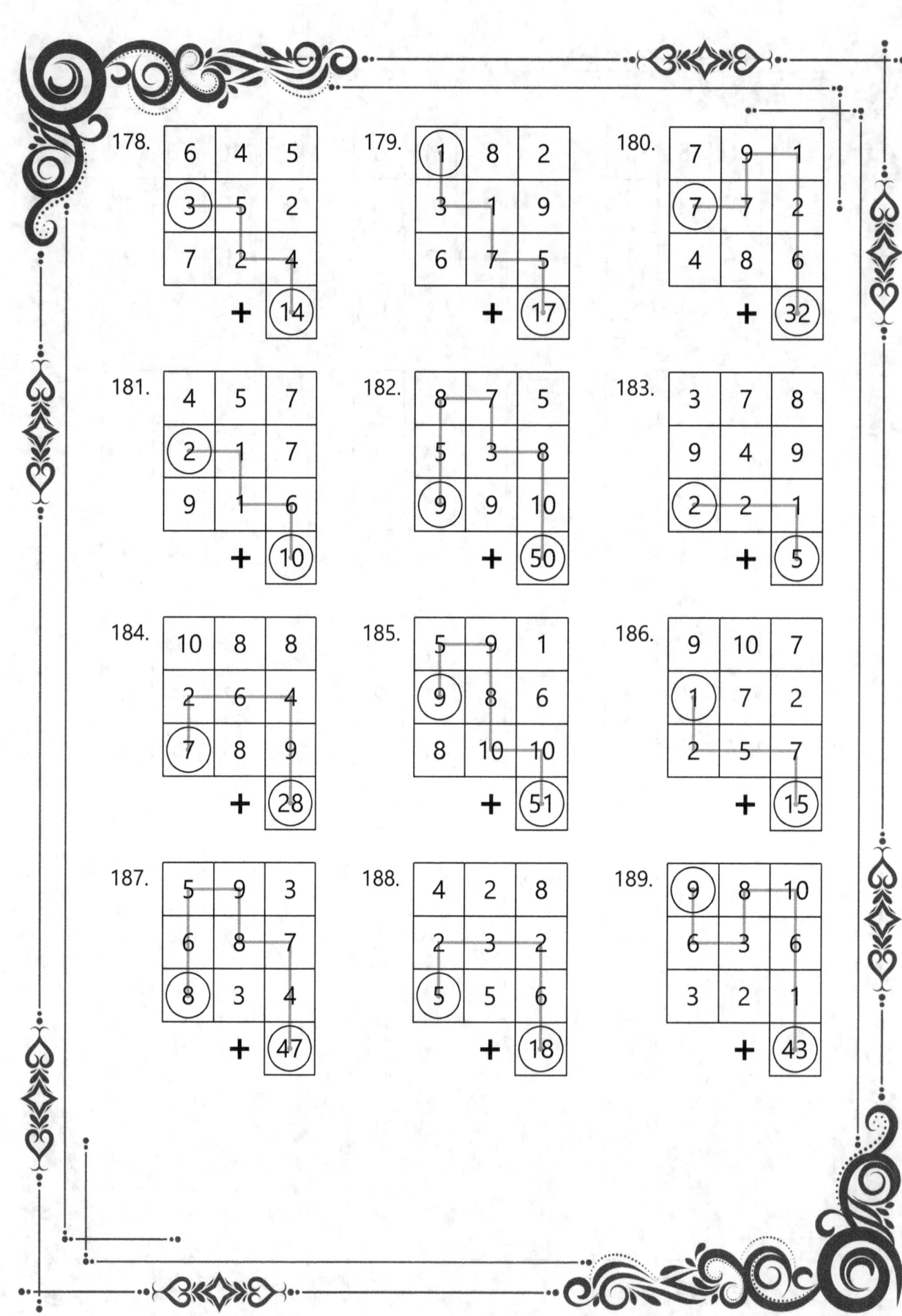

Secret Trails Subtraction

190.

46	9	4
66	74	55
(363)	90	51

− (93)

191.

36	67	85
22	30	16
(348)	45	63

− (59)

192.

(283)	23	9
53	95	82
33	65	94

− (38)

193.

41	77	57
(353)	52	69
39	96	25

− (72)

194.

46	38	40
(329)	99	63
1	59	67

− (75)

195.

22	29	97
(299)	27	60
11	14	71

− (15)

196.

78	63	21
(205)	58	1
84	35	54

− (92)

197.

82	74	99
(368)	62	1
53	85	68

− (99)

198.

69	87	25
(261)	50	18
42	25	48

− (33)

199.

53	87	89
(71)	18	12
20	33	15

− (5)

200.

41	66	98
39	11	48
(233)	98	54

− (31)

201.

(278)	26	47
43	94	91
39	88	42

− (8)

202.

60	59	87
(399)	80	95
86	94	82

− (16)

203.

(351)	20	71
65	88	3
90	86	22

− (82)

204.

88	44	72
(182)	32	27
41	47	12

− (23)

205.

(292)	33	91
75	44	82
80	83	60

− (73)

206.

63	38	17
4	45	88
(343)	82	55

− (56)

207.

2	54	66
(194)	53	17
87	65	84

− (40)

208.

55	39	82
(326)	78	50
29	97	22

− (55)

209.

24	7	26
(272)	40	51
27	69	79

− (6)

210.

72	60	93
(241)	25	20
67	52	22

− (21)

211.

95	15	99
(239)	63	24
56	62	27

− (7)

212.

59	60	60
(423)	68	95
28	60	100

− (49)

213.

(304)	19	86
40	82	96
47	62	30

− (90)

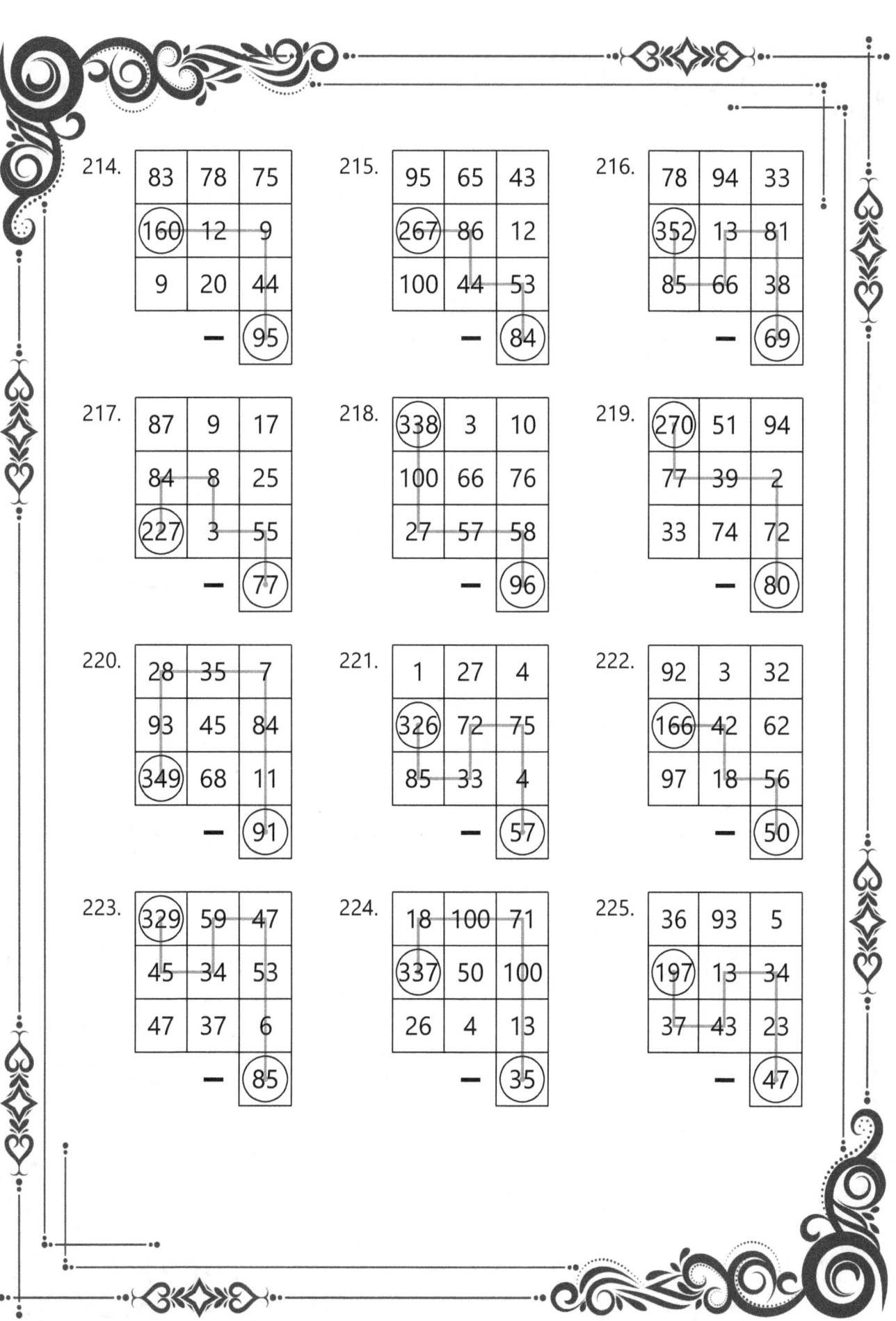

214.

83	78	75
(160)	12	9
9	20	44
	—	(95)

215.

95	65	43
(267)	86	12
100	44	53
	—	(84)

216.

78	94	33
(352)	13	81
85	66	38
	—	(69)

217.

87	9	17
84	8	25
(227)	3	55
	—	(77)

218.

(338)	3	10
100	66	76
27	57	58
	—	(96)

219.

(270)	51	94
77	39	2
33	74	72
	—	(80)

220.

28	35	7
93	45	84
(349)	68	11
	—	(91)

221.

1	27	4
(326)	72	75
85	33	4
	—	(57)

222.

92	3	32
(166)	42	62
97	18	56
	—	(50)

223.

(329)	59	47
45	34	53
47	37	6
	—	(85)

224.

18	100	71
(337)	50	100
26	4	13
	—	(35)

225.

36	93	5
(197)	13	34
37	43	23
	—	(47)

226.

58	93	22
97	86	51
(487)	48	83

− (22)

227.

18	62	35
85	19	91
(184)	41	15

− (18)

228.

60	40	42
(300)	71	98
57	38	97

− (34)

229.

17	96	29
(239)	71	58
39	50	24

− (94)

230.

89	91	38
(356)	95	79
21	2	14

− (65)

231.

28	42	4
(278)	3	64
11	48	63

− (89)

232.

59	8	15
(319)	51	80
48	50	53

− (98)

233.

73	32	3
(346)	73	47
88	50	63

− (25)

234.

(233)	74	39
1	64	27
67	47	25

− (68)

235.

34	36	55
(350)	46	70
35	98	100

− (1)

236.

13	58	13
(238)	40	27
5	90	26

− (74)

237.

93	98	88
23	79	81
(246)	6	43

− (20)

Sudoku Easy Level

Fill the grid so that every row, every column and every 3x3 box contains the numbers 1 to 9.

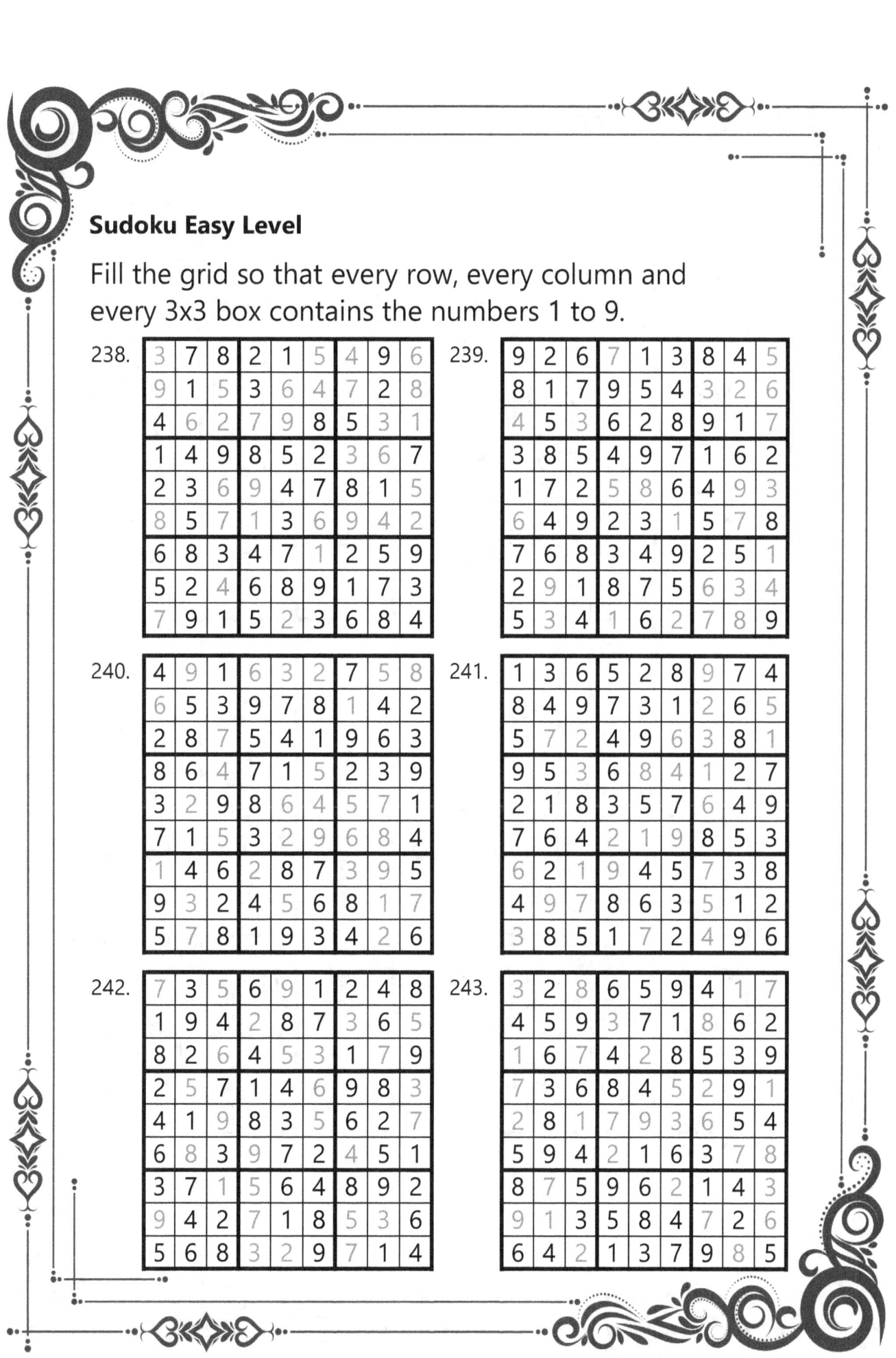

238.

3	7	8	2	1	5	4	9	6
9	1	5	3	6	4	7	2	8
4	6	2	7	9	8	5	3	1
1	4	9	8	5	2	3	6	7
2	3	6	9	4	7	8	1	5
8	5	7	1	3	6	9	4	2
6	8	3	4	7	1	2	5	9
5	2	4	6	8	9	1	7	3
7	9	1	5	2	3	6	8	4

239.

9	2	6	7	1	3	8	4	5
8	1	7	9	5	4	3	2	6
4	5	3	6	2	8	9	1	7
3	8	5	4	9	7	1	6	2
1	7	2	5	8	6	4	9	3
6	4	9	2	3	1	5	7	8
7	6	8	3	4	9	2	5	1
2	9	1	8	7	5	6	3	4
5	3	4	1	6	2	7	8	9

240.

4	9	1	6	3	2	7	5	8
6	5	3	9	7	8	1	4	2
2	8	7	5	4	1	9	6	3
8	6	4	7	1	5	2	3	9
3	2	9	8	6	4	5	7	1
7	1	5	3	2	9	6	8	4
1	4	6	2	8	7	3	9	5
9	3	2	4	5	6	8	1	7
5	7	8	1	9	3	4	2	6

241.

1	3	6	5	2	8	9	7	4
8	4	9	7	3	1	2	6	5
5	7	2	4	9	6	3	8	1
9	5	3	6	8	4	1	2	7
2	1	8	3	5	7	6	4	9
7	6	4	2	1	9	8	5	3
6	2	1	9	4	5	7	3	8
4	9	7	8	6	3	5	1	2
3	8	5	1	7	2	4	9	6

242.

7	3	5	6	9	1	2	4	8
1	9	4	2	8	7	3	6	5
8	2	6	4	5	3	1	7	9
2	5	7	1	4	6	9	8	3
4	1	9	8	3	5	6	2	7
6	8	3	9	7	2	4	5	1
3	7	1	5	6	4	8	9	2
9	4	2	7	1	8	5	3	6
5	6	8	3	2	9	7	1	4

243.

3	2	8	6	5	9	4	1	7
4	5	9	3	7	1	8	6	2
1	6	7	4	2	8	5	3	9
7	3	6	8	4	5	2	9	1
2	8	1	7	9	3	6	5	4
5	9	4	2	1	6	3	7	8
8	7	5	9	6	2	1	4	3
9	1	3	5	8	4	7	2	6
6	4	2	1	3	7	9	8	5

244.

2	5	9	1	4	8	7	6	3
1	7	3	9	5	6	8	2	4
6	4	8	3	2	7	5	9	1
3	1	4	2	9	5	6	7	8
8	9	2	7	6	1	4	3	5
5	6	7	8	3	4	9	1	2
4	2	6	5	7	3	1	8	9
7	3	1	4	8	9	2	5	6
9	8	5	6	1	2	3	4	7

245.

9	7	2	6	5	3	1	4	8
3	1	6	8	7	4	5	9	2
5	8	4	9	2	1	3	7	6
8	4	1	5	3	9	2	6	7
2	5	9	1	6	7	4	8	3
6	3	7	2	4	8	9	1	5
4	2	3	7	1	6	8	5	9
7	9	5	4	8	2	6	3	1
1	6	8	3	9	5	7	2	4

246.

4	2	9	3	8	7	6	1	5
7	3	6	2	5	1	9	4	8
8	5	1	6	9	4	2	3	7
1	8	4	9	3	2	7	5	6
5	9	7	4	1	6	8	2	3
2	6	3	5	7	8	4	9	1
3	4	2	8	6	5	1	7	9
6	7	5	1	2	9	3	8	4
9	1	8	7	4	3	5	6	2

247.

4	5	7	3	9	2	1	6	8
3	6	2	7	1	8	5	9	4
9	1	8	5	4	6	7	3	2
8	2	1	9	6	5	3	4	7
5	3	4	8	2	7	9	1	6
7	9	6	1	3	4	8	2	5
2	7	9	4	8	1	6	5	3
6	8	3	2	5	9	4	7	1
1	4	5	6	7	3	2	8	9

248.

1	7	5	9	4	2	6	8	3
9	6	3	1	5	8	4	2	7
8	4	2	6	3	7	1	5	9
6	5	9	2	7	3	8	1	4
4	3	8	5	9	1	2	7	6
2	1	7	4	8	6	3	9	5
5	8	4	3	1	9	7	6	2
3	2	1	7	6	5	9	4	8
7	9	6	8	2	4	5	3	1

249.

3	2	5	9	7	6	1	4	8
6	8	7	4	5	1	9	3	2
4	1	9	2	3	8	7	5	6
8	4	2	5	6	7	3	1	9
9	3	1	8	2	4	6	7	5
5	7	6	1	9	3	2	8	4
2	9	4	3	1	5	8	6	7
7	5	3	6	8	2	4	9	1
1	6	8	7	4	9	5	2	3

250.

4	1	6	9	2	5	3	8	7
8	2	3	4	7	1	5	6	9
7	9	5	6	8	3	1	2	4
3	8	7	5	6	4	9	1	2
5	6	1	8	9	2	4	7	3
2	4	9	1	3	7	8	5	6
6	5	4	2	1	9	7	3	8
9	7	8	3	5	6	2	4	1
1	3	2	7	4	8	6	9	5

251.

8	4	2	3	6	1	7	5	9
5	3	1	7	4	9	8	2	6
7	9	6	5	2	8	1	4	3
2	6	5	9	1	7	4	3	8
1	7	9	4	8	3	5	6	2
4	8	3	2	5	6	9	7	1
6	5	8	1	7	2	3	9	4
9	1	4	6	3	5	2	8	7
3	2	7	8	9	4	6	1	5

252.

4	7	3	9	1	2	5	6	8
6	5	8	4	3	7	9	2	1
2	9	1	8	5	6	3	7	4
3	4	2	6	8	1	7	5	9
7	6	9	2	4	5	8	1	3
1	8	5	7	9	3	2	4	6
9	3	6	5	7	4	1	8	2
8	2	7	1	6	9	4	3	5
5	1	4	3	2	8	6	9	7

253.

7	6	9	3	2	4	8	1	5
1	8	5	7	9	6	3	2	4
3	2	4	8	5	1	6	7	9
2	7	1	5	6	3	9	4	8
6	4	3	2	8	9	7	5	1
9	5	8	1	4	7	2	6	3
5	1	7	6	3	8	4	9	2
4	3	2	9	7	5	1	8	6
8	9	6	4	1	2	5	3	7

254.

4	3	8	6	2	1	7	9	5
7	5	9	8	4	3	1	6	2
6	1	2	7	9	5	4	3	8
8	9	5	3	7	6	2	1	4
3	6	1	2	5	4	8	7	9
2	7	4	1	8	9	3	5	6
9	4	3	5	1	2	6	8	7
5	8	6	4	3	7	9	2	1
1	2	7	9	6	8	5	4	3

255.

5	6	3	4	7	2	8	1	9
9	8	1	5	6	3	7	2	4
2	7	4	1	9	8	3	5	6
4	3	2	8	1	9	5	6	7
1	9	8	7	5	6	2	4	3
6	5	7	2	3	4	1	9	8
7	2	6	3	4	1	9	8	5
3	1	9	6	8	5	4	7	2
8	4	5	9	2	7	6	3	1

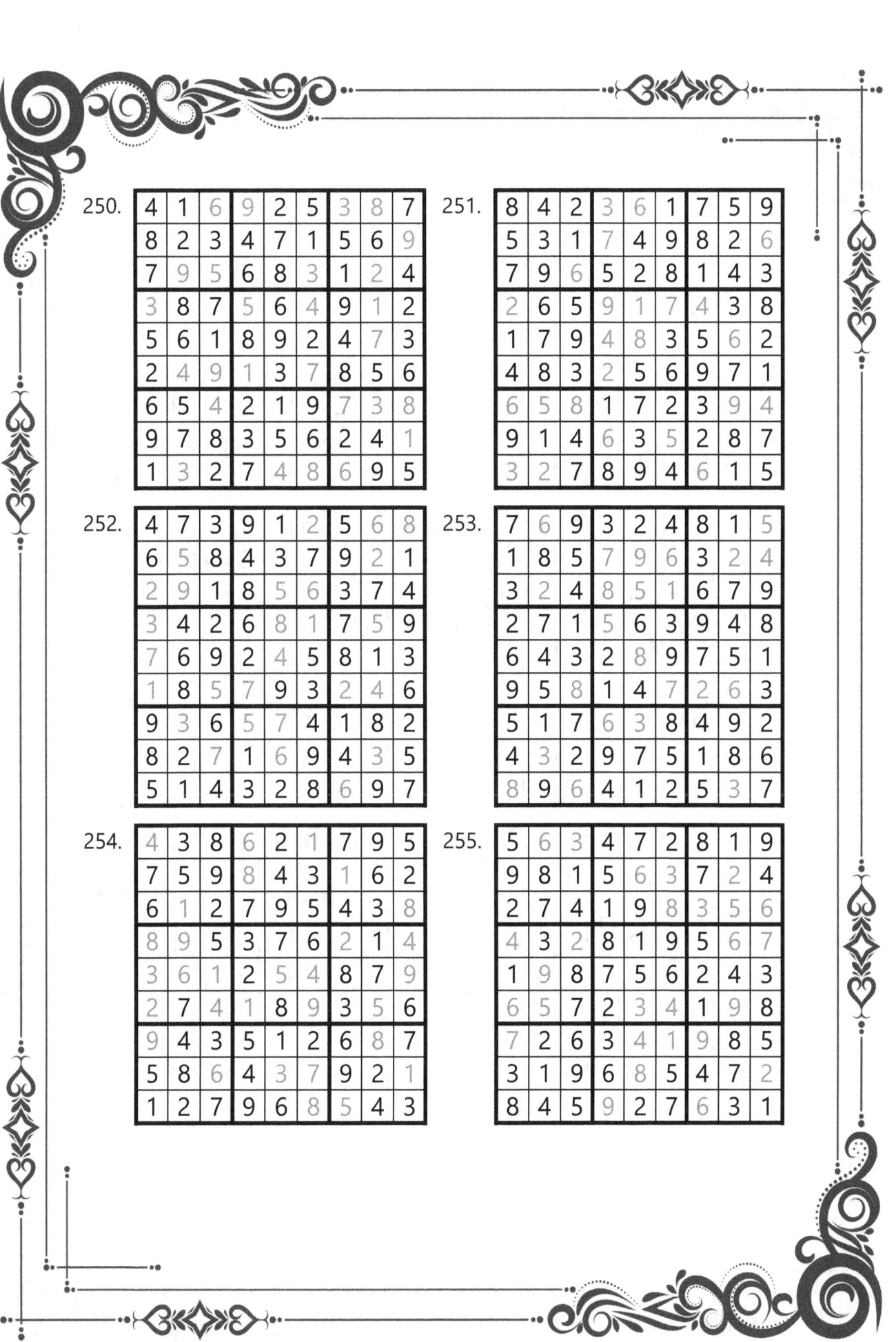

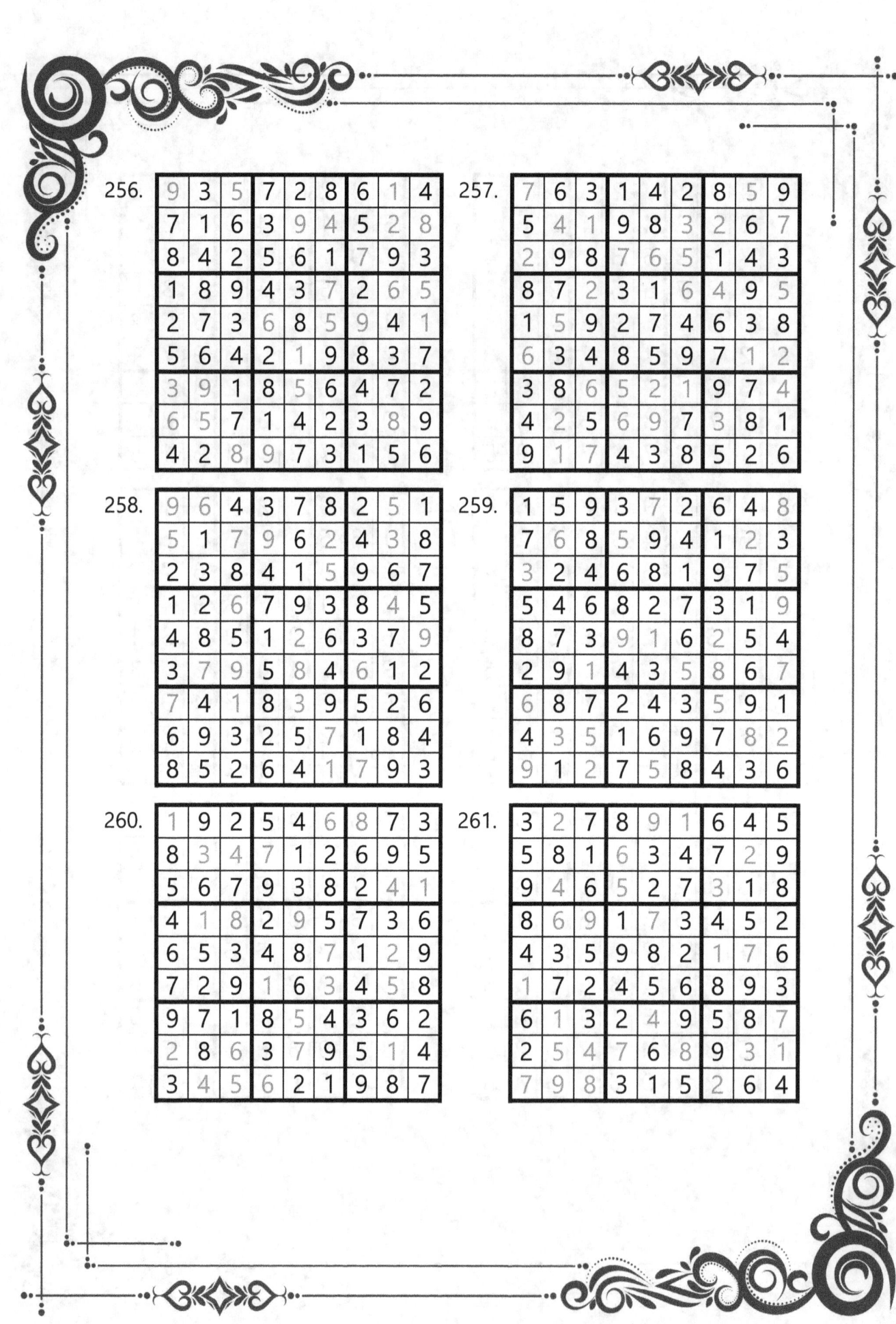

256.

9	3	5	7	2	8	6	1	4
7	1	6	3	9	4	5	2	8
8	4	2	5	6	1	7	9	3
1	8	9	4	3	7	2	6	5
2	7	3	6	8	5	9	4	1
5	6	4	2	1	9	8	3	7
3	9	1	8	5	6	4	7	2
6	5	7	1	4	2	3	8	9
4	2	8	9	7	3	1	5	6

257.

7	6	3	1	4	2	8	5	9
5	4	1	9	8	3	2	6	7
2	9	8	7	6	5	1	4	3
8	7	2	3	1	6	4	9	5
1	5	9	2	7	4	6	3	8
6	3	4	8	5	9	7	1	2
3	8	6	5	2	1	9	7	4
4	2	5	6	9	7	3	8	1
9	1	7	4	3	8	5	2	6

258.

9	6	4	3	7	8	2	5	1
5	1	7	9	6	2	4	3	8
2	3	8	4	1	5	9	6	7
1	2	6	7	9	3	8	4	5
4	8	5	1	2	6	3	7	9
3	7	9	5	8	4	6	1	2
7	4	1	8	3	9	5	2	6
6	9	3	2	5	7	1	8	4
8	5	2	6	4	1	7	9	3

259.

1	5	9	3	7	2	6	4	8
7	6	8	5	9	4	1	2	3
3	2	4	6	8	1	9	7	5
5	4	6	8	2	7	3	1	9
8	7	3	9	1	6	2	5	4
2	9	1	4	3	5	8	6	7
6	8	7	2	4	3	5	9	1
4	3	5	1	6	9	7	8	2
9	1	2	7	5	8	4	3	6

260.

1	9	2	5	4	6	8	7	3
8	3	4	7	1	2	6	9	5
5	6	7	9	3	8	2	4	1
4	1	8	2	9	5	7	3	6
6	5	3	4	8	7	1	2	9
7	2	9	1	6	3	4	5	8
9	7	1	8	5	4	3	6	2
2	8	6	3	7	9	5	1	4
3	4	5	6	2	1	9	8	7

261.

3	2	7	8	9	1	6	4	5
5	8	1	6	3	4	7	2	9
9	4	6	5	2	7	3	1	8
8	6	9	1	7	3	4	5	2
4	3	5	9	8	2	1	7	6
1	7	2	4	5	6	8	9	3
6	1	3	2	4	9	5	8	7
2	5	4	7	6	8	9	3	1
7	9	8	3	1	5	2	6	4

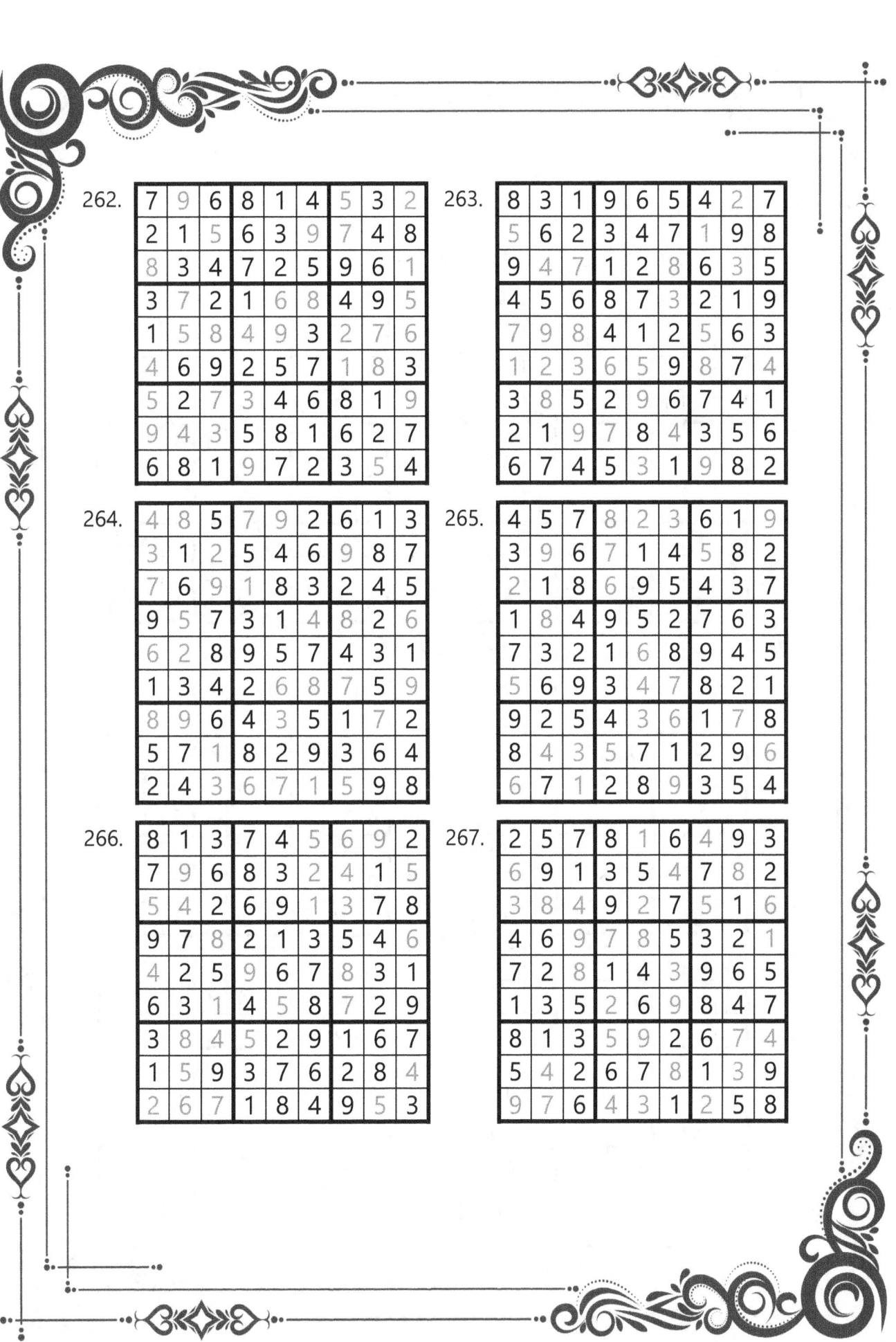

262.

7	9	6	8	1	4	5	3	2
2	1	5	6	3	9	7	4	8
8	3	4	7	2	5	9	6	1
3	7	2	1	6	8	4	9	5
1	5	8	4	9	3	2	7	6
4	6	9	2	5	7	1	8	3
5	2	7	3	4	6	8	1	9
9	4	3	5	8	1	6	2	7
6	8	1	9	7	2	3	5	4

263.

8	3	1	9	6	5	4	2	7
5	6	2	3	4	7	1	9	8
9	4	7	1	2	8	6	3	5
4	5	6	8	7	3	2	1	9
7	9	8	4	1	2	5	6	3
1	2	3	6	5	9	8	7	4
3	8	5	2	9	6	7	4	1
2	1	9	7	8	4	3	5	6
6	7	4	5	3	1	9	8	2

264.

4	8	5	7	9	2	6	1	3
3	1	2	5	4	6	9	8	7
7	6	9	1	8	3	2	4	5
9	5	7	3	1	4	8	2	6
6	2	8	9	5	7	4	3	1
1	3	4	2	6	8	7	5	9
8	9	6	4	3	5	1	7	2
5	7	1	8	2	9	3	6	4
2	4	3	6	7	1	5	9	8

265.

4	5	7	8	2	3	6	1	9
3	9	6	7	1	4	5	8	2
2	1	8	6	9	5	4	3	7
1	8	4	9	5	2	7	6	3
7	3	2	1	6	8	9	4	5
5	6	9	3	4	7	8	2	1
9	2	5	4	3	6	1	7	8
8	4	3	5	7	1	2	9	6
6	7	1	2	8	9	3	5	4

266.

8	1	3	7	4	5	6	9	2
7	9	6	8	3	2	4	1	5
5	4	2	6	9	1	3	7	8
9	7	8	2	1	3	5	4	6
4	2	5	9	6	7	8	3	1
6	3	1	4	5	8	7	2	9
3	8	4	5	2	9	1	6	7
1	5	9	3	7	6	2	8	4
2	6	7	1	8	4	9	5	3

267.

2	5	7	8	1	6	4	9	3
6	9	1	3	5	4	7	8	2
3	8	4	9	2	7	5	1	6
4	6	9	7	8	5	3	2	1
7	2	8	1	4	3	9	6	5
1	3	5	2	6	9	8	4	7
8	1	3	5	9	2	6	7	4
5	4	2	6	7	8	1	3	9
9	7	6	4	3	1	2	5	8

Sudoku Medium Level

Fill the grid so that every row, every column and every 3x3 box contains the numbers 1 to 9.

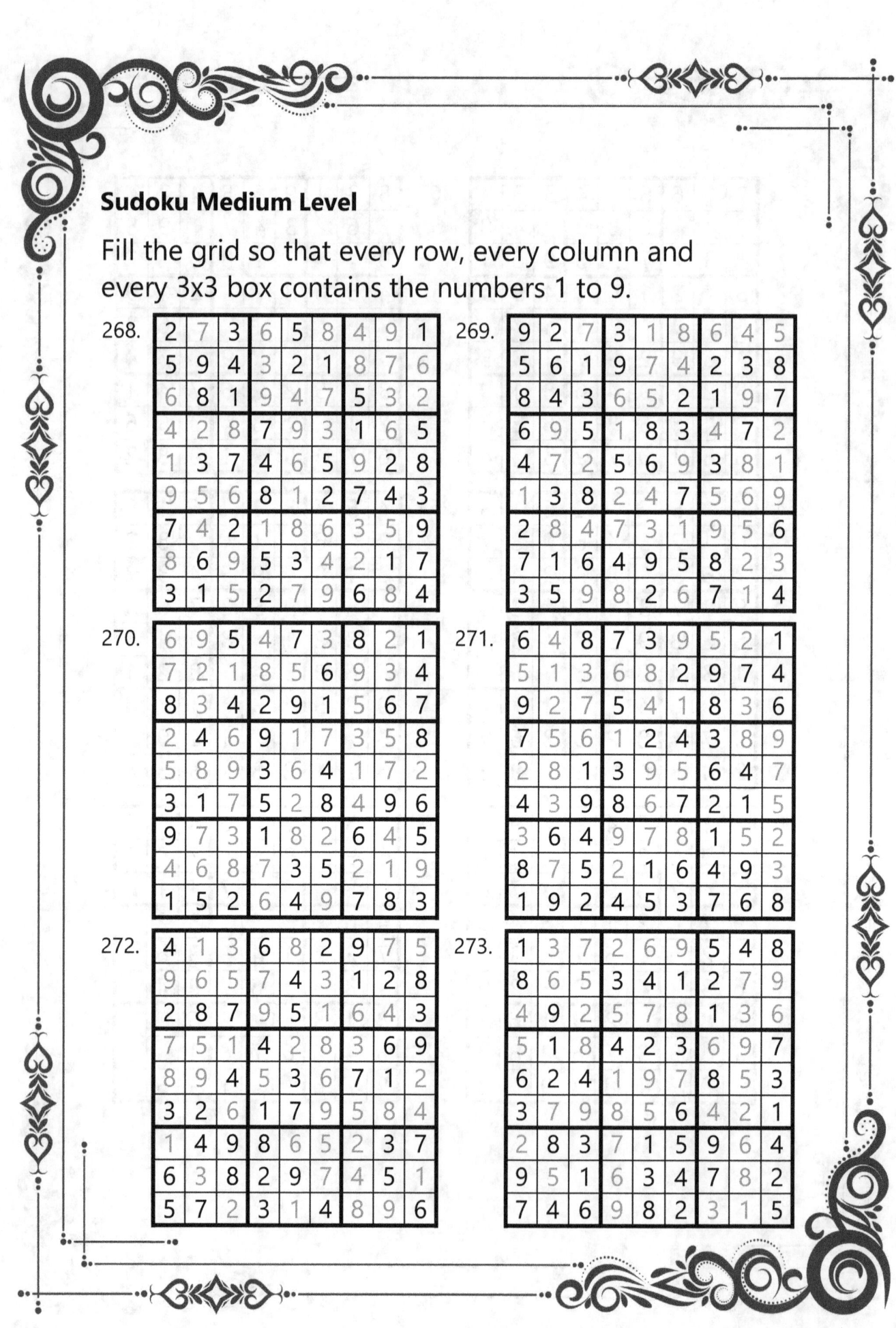

268.

2	7	3	6	5	8	4	9	1
5	9	4	3	2	1	8	7	6
6	8	1	9	4	7	5	3	2
4	2	8	7	9	3	1	6	5
1	3	7	4	6	5	9	2	8
9	5	6	8	1	2	7	4	3
7	4	2	1	8	6	3	5	9
8	6	9	5	3	4	2	1	7
3	1	5	2	7	9	6	8	4

269.

9	2	7	3	1	8	6	4	5
5	6	1	9	7	4	2	3	8
8	4	3	6	5	2	1	9	7
6	9	5	1	8	3	4	7	2
4	7	2	5	6	9	3	8	1
1	3	8	2	4	7	5	6	9
2	8	4	7	3	1	9	5	6
7	1	6	4	9	5	8	2	3
3	5	9	8	2	6	7	1	4

270.

6	9	5	4	7	3	8	2	1
7	2	1	8	5	6	9	3	4
8	3	4	2	9	1	5	6	7
2	4	6	9	1	7	3	5	8
5	8	9	3	6	4	1	7	2
3	1	7	5	2	8	4	9	6
9	7	3	1	8	2	6	4	5
4	6	8	7	3	5	2	1	9
1	5	2	6	4	9	7	8	3

271.

6	4	8	7	3	9	5	2	1
5	1	3	6	8	2	9	7	4
9	2	7	5	4	1	8	3	6
7	5	6	1	2	4	3	8	9
2	8	1	3	9	5	6	4	7
4	3	9	8	6	7	2	1	5
3	6	4	9	7	8	1	5	2
8	7	5	2	1	6	4	9	3
1	9	2	4	5	3	7	6	8

272.

4	1	3	6	8	2	9	7	5
9	6	5	7	4	3	1	2	8
2	8	7	9	5	1	6	4	3
7	5	1	4	2	8	3	6	9
8	9	4	5	3	6	7	1	2
3	2	6	1	7	9	5	8	4
1	4	9	8	6	5	2	3	7
6	3	8	2	9	7	4	5	1
5	7	2	3	1	4	8	9	6

273.

1	3	7	2	6	9	5	4	8
8	6	5	3	4	1	2	7	9
4	9	2	5	7	8	1	3	6
5	1	8	4	2	3	6	9	7
6	2	4	1	9	7	8	5	3
3	7	9	8	5	6	4	2	1
2	8	3	7	1	5	9	6	4
9	5	1	6	3	4	7	8	2
7	4	6	9	8	2	3	1	5

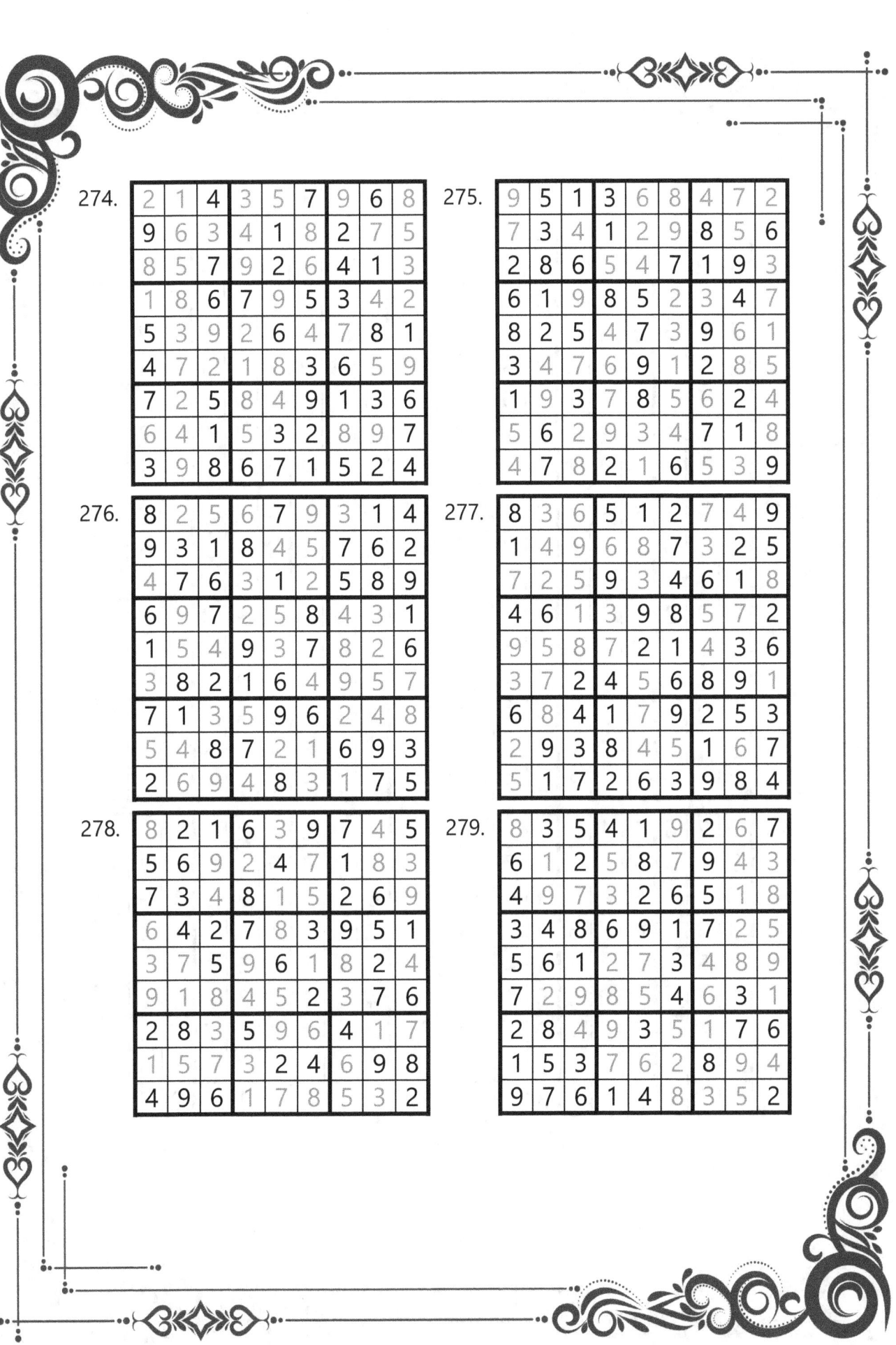

274.

2	1	4	3	5	7	9	6	8
9	6	3	4	1	8	2	7	5
8	5	7	9	2	6	4	1	3
1	8	6	7	9	5	3	4	2
5	3	9	2	6	4	7	8	1
4	7	2	1	8	3	6	5	9
7	2	5	8	4	9	1	3	6
6	4	1	5	3	2	8	9	7
3	9	8	6	7	1	5	2	4

275.

9	5	1	3	6	8	4	7	2
7	3	4	1	2	9	8	5	6
2	8	6	5	4	7	1	9	3
6	1	9	8	5	2	3	4	7
8	2	5	4	7	3	9	6	1
3	4	7	6	9	1	2	8	5
1	9	3	7	8	5	6	2	4
5	6	2	9	3	4	7	1	8
4	7	8	2	1	6	5	3	9

276.

8	2	5	6	7	9	3	1	4
9	3	1	8	4	5	7	6	2
4	7	6	3	1	2	5	8	9
6	9	7	2	5	8	4	3	1
1	5	4	9	3	7	8	2	6
3	8	2	1	6	4	9	5	7
7	1	3	5	9	6	2	4	8
5	4	8	7	2	1	6	9	3
2	6	9	4	8	3	1	7	5

277.

8	3	6	5	1	2	7	4	9
1	4	9	6	8	7	3	2	5
7	2	5	9	3	4	6	1	8
4	6	1	3	9	8	5	7	2
9	5	8	7	2	1	4	3	6
3	7	2	4	5	6	8	9	1
6	8	4	1	7	9	2	5	3
2	9	3	8	4	5	1	6	7
5	1	7	2	6	3	9	8	4

278.

8	2	1	6	3	9	7	4	5
5	6	9	2	4	7	1	8	3
7	3	4	8	1	5	2	6	9
6	4	2	7	8	3	9	5	1
3	7	5	9	6	1	8	2	4
9	1	8	4	5	2	3	7	6
2	8	3	5	9	6	4	1	7
1	5	7	3	2	4	6	9	8
4	9	6	1	7	8	5	3	2

279.

8	3	5	4	1	9	2	6	7
6	1	2	5	8	7	9	4	3
4	9	7	3	2	6	5	1	8
3	4	8	6	9	1	7	2	5
5	6	1	2	7	3	4	8	9
7	2	9	8	5	4	6	3	1
2	8	4	9	3	5	1	7	6
1	5	3	7	6	2	8	9	4
9	7	6	1	4	8	3	5	2

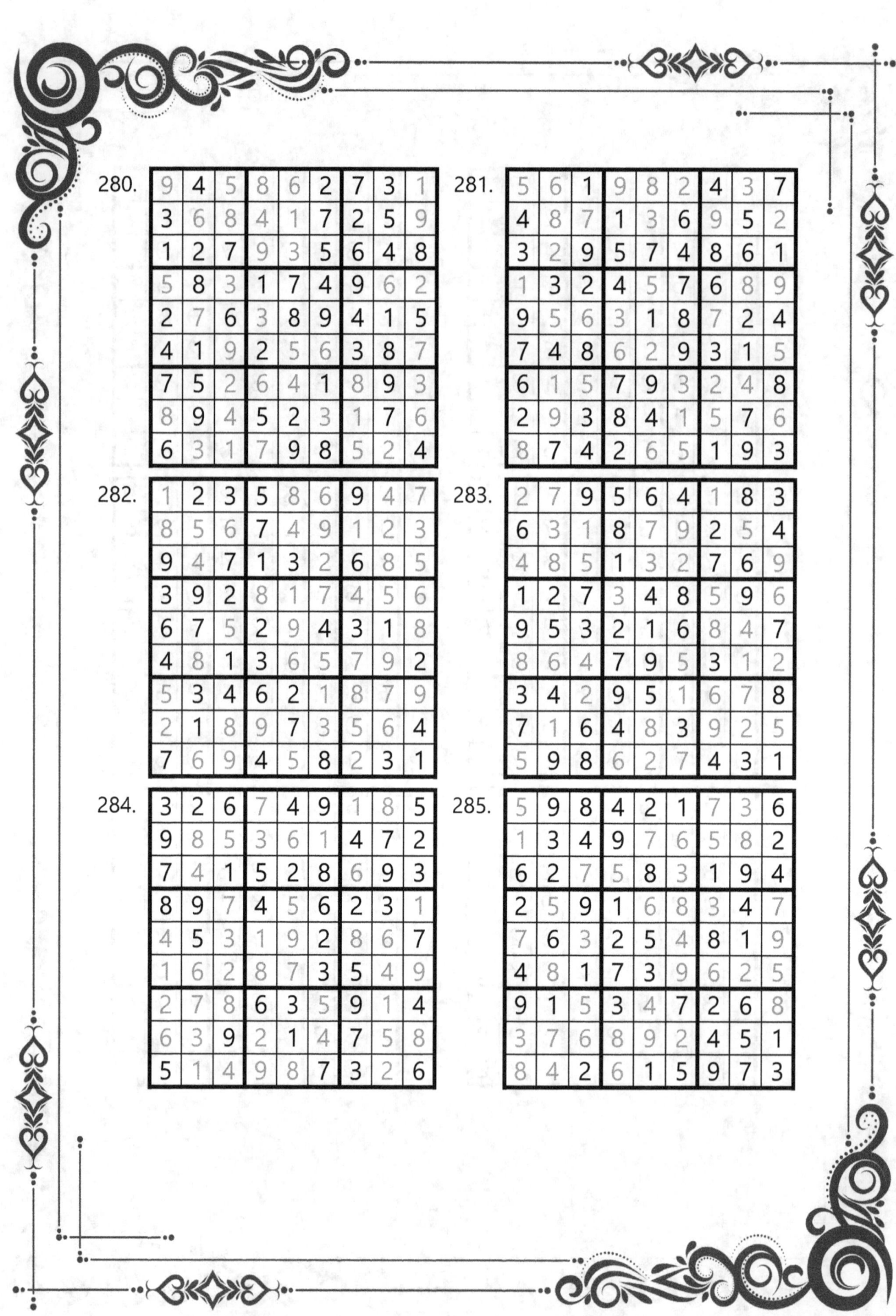

280.

9	4	5	8	6	2	7	3	1
3	6	8	4	1	7	2	5	9
1	2	7	9	3	5	6	4	8
5	8	3	1	7	4	9	6	2
2	7	6	3	8	9	4	1	5
4	1	9	2	5	6	3	8	7
7	5	2	6	4	1	8	9	3
8	9	4	5	2	3	1	7	6
6	3	1	7	9	8	5	2	4

281.

5	6	1	9	8	2	4	3	7
4	8	7	1	3	6	9	5	2
3	2	9	5	7	4	8	6	1
1	3	2	4	5	7	6	8	9
9	5	6	3	1	8	7	2	4
7	4	8	6	2	9	3	1	5
6	1	5	7	9	3	2	4	8
2	9	3	8	4	1	5	7	6
8	7	4	2	6	5	1	9	3

282.

1	2	3	5	8	6	9	4	7
8	5	6	7	4	9	1	2	3
9	4	7	1	3	2	6	8	5
3	9	2	8	1	7	4	5	6
6	7	5	2	9	4	3	1	8
4	8	1	3	6	5	7	9	2
5	3	4	6	2	1	8	7	9
2	1	8	9	7	3	5	6	4
7	6	9	4	5	8	2	3	1

283.

2	7	9	5	6	4	1	8	3
6	3	1	8	7	9	2	5	4
4	8	5	1	3	2	7	6	9
1	2	7	3	4	8	5	9	6
9	5	3	2	1	6	8	4	7
8	6	4	7	9	5	3	1	2
3	4	2	9	5	1	6	7	8
7	1	6	4	8	3	9	2	5
5	9	8	6	2	7	4	3	1

284.

3	2	6	7	4	9	1	8	5
9	8	5	3	6	1	4	7	2
7	4	1	5	2	8	6	9	3
8	9	7	4	5	6	2	3	1
4	5	3	1	9	2	8	6	7
1	6	2	8	7	3	5	4	9
2	7	8	6	3	5	9	1	4
6	3	9	2	1	4	7	5	8
5	1	4	9	8	7	3	2	6

285.

5	9	8	4	2	1	7	3	6
1	3	4	9	7	6	5	8	2
6	2	7	5	8	3	1	9	4
2	5	9	1	6	8	3	4	7
7	6	3	2	5	4	8	1	9
4	8	1	7	3	9	6	2	5
9	1	5	3	4	7	2	6	8
3	7	6	8	9	2	4	5	1
8	4	2	6	1	5	9	7	3

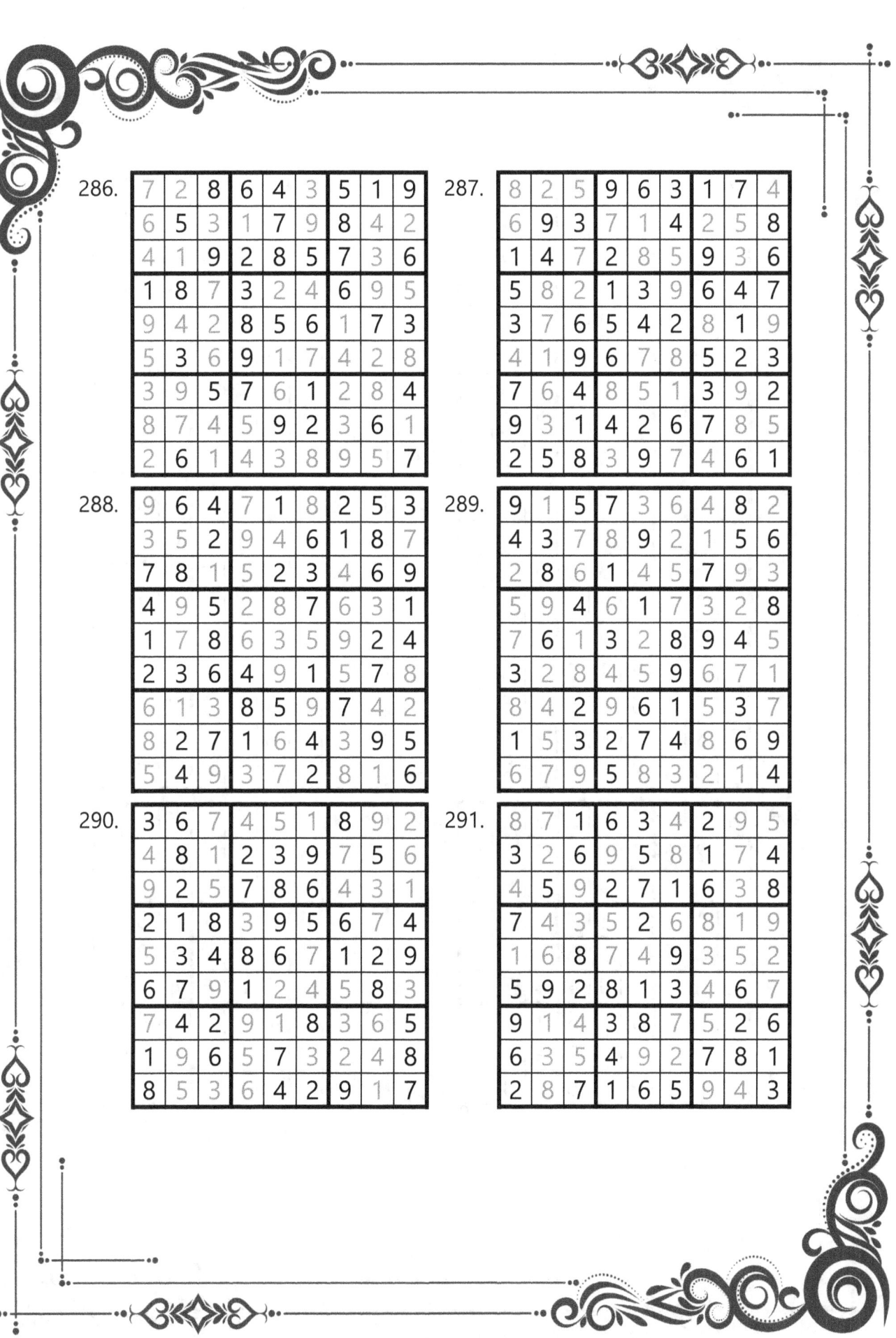

286.

7	2	8	6	4	3	5	1	9
6	5	3	1	7	9	8	4	2
4	1	9	2	8	5	7	3	6
1	8	7	3	2	4	6	9	5
9	4	2	8	5	6	1	7	3
5	3	6	9	1	7	4	2	8
3	9	5	7	6	1	2	8	4
8	7	4	5	9	2	3	6	1
2	6	1	4	3	8	9	5	7

287.

8	2	5	9	6	3	1	7	4
6	9	3	7	1	4	2	5	8
1	4	7	2	8	5	9	3	6
5	8	2	1	3	9	6	4	7
3	7	6	5	4	2	8	1	9
4	1	9	6	7	8	5	2	3
7	6	4	8	5	1	3	9	2
9	3	1	4	2	6	7	8	5
2	5	8	3	9	7	4	6	1

288.

9	6	4	7	1	8	2	5	3
3	5	2	9	4	6	1	8	7
7	8	1	5	2	3	4	6	9
4	9	5	2	8	7	6	3	1
1	7	8	6	3	5	9	2	4
2	3	6	4	9	1	5	7	8
6	1	3	8	5	9	7	4	2
8	2	7	1	6	4	3	9	5
5	4	9	3	7	2	8	1	6

289.

9	1	5	7	3	6	4	8	2
4	3	7	8	9	2	1	5	6
2	8	6	1	4	5	7	9	3
5	9	4	6	1	7	3	2	8
7	6	1	3	2	8	9	4	5
3	2	8	4	5	9	6	7	1
8	4	2	9	6	1	5	3	7
1	5	3	2	7	4	8	6	9
6	7	9	5	8	3	2	1	4

290.

3	6	7	4	5	1	8	9	2
4	8	1	2	3	9	7	5	6
9	2	5	7	8	6	4	3	1
2	1	8	3	9	5	6	7	4
5	3	4	8	6	7	1	2	9
6	7	9	1	2	4	5	8	3
7	4	2	9	1	8	3	6	5
1	9	6	5	7	3	2	4	8
8	5	3	6	4	2	9	1	7

291.

8	7	1	6	3	4	2	9	5
3	2	6	9	5	8	1	7	4
4	5	9	2	7	1	6	3	8
7	4	3	5	2	6	8	1	9
1	6	8	7	4	9	3	5	2
5	9	2	8	1	3	4	6	7
9	1	4	3	8	7	5	2	6
6	3	5	4	9	2	7	8	1
2	8	7	1	6	5	9	4	3

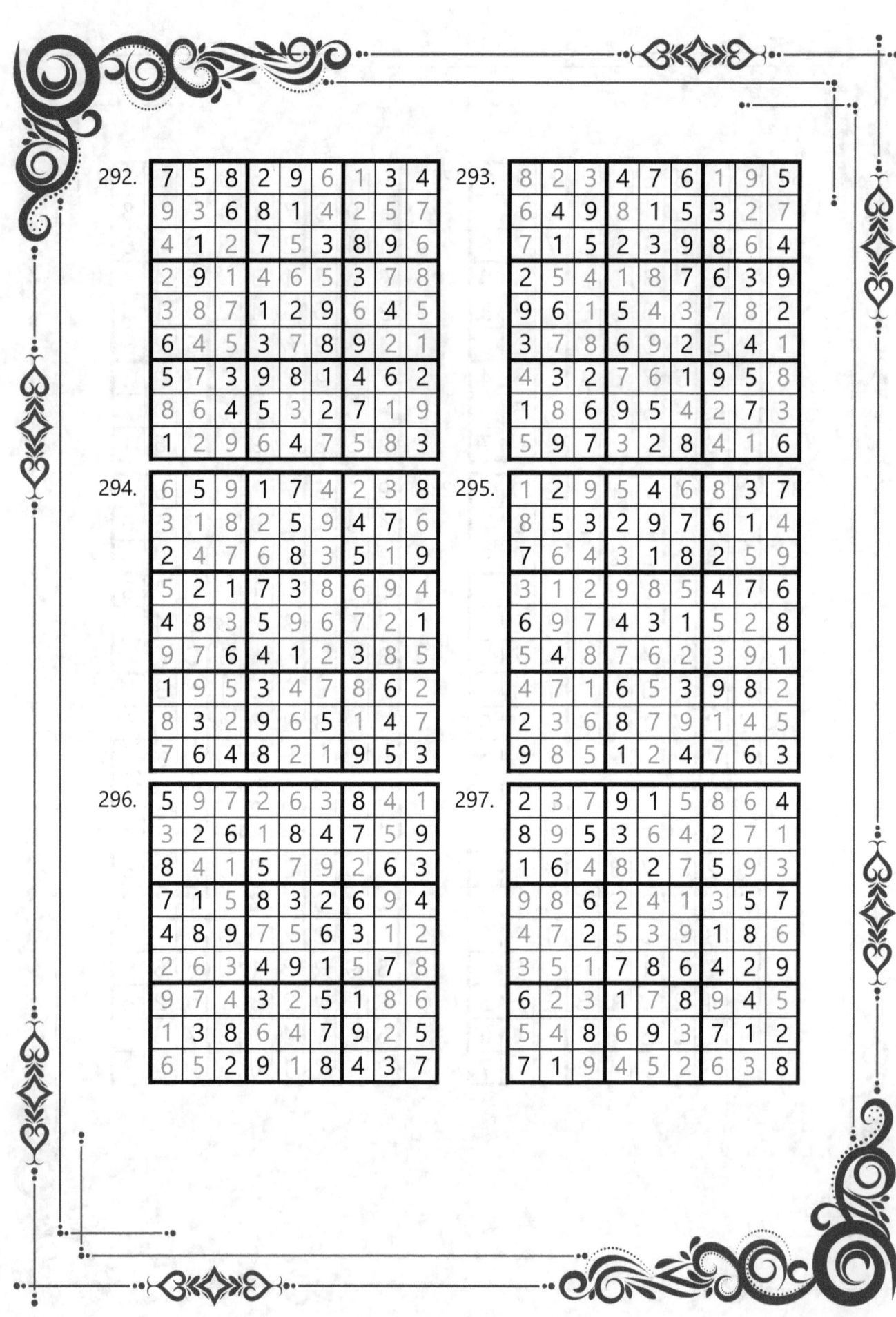

292.

7	5	8	2	9	6	1	3	4
9	3	6	8	1	4	2	5	7
4	1	2	7	5	3	8	9	6
2	9	1	4	6	5	3	7	8
3	8	7	1	2	9	6	4	5
6	4	5	3	7	8	9	2	1
5	7	3	9	8	1	4	6	2
8	6	4	5	3	2	7	1	9
1	2	9	6	4	7	5	8	3

293.

8	2	3	4	7	6	1	9	5
6	4	9	8	1	5	3	2	7
7	1	5	2	3	9	8	6	4
2	5	4	1	8	7	6	3	9
9	6	1	5	4	3	7	8	2
3	7	8	6	9	2	5	4	1
4	3	2	7	6	1	9	5	8
1	8	6	9	5	4	2	7	3
5	9	7	3	2	8	4	1	6

294.

6	5	9	1	7	4	2	3	8
3	1	8	2	5	9	4	7	6
2	4	7	6	8	3	5	1	9
5	2	1	7	3	8	6	9	4
4	8	3	5	9	6	7	2	1
9	7	6	4	1	2	3	8	5
1	9	5	3	4	7	8	6	2
8	3	2	9	6	5	1	4	7
7	6	4	8	2	1	9	5	3

295.

1	2	9	5	4	6	8	3	7
8	5	3	2	9	7	6	1	4
7	6	4	3	1	8	2	5	9
3	1	2	9	8	5	4	7	6
6	9	7	4	3	1	5	2	8
5	4	8	7	6	2	3	9	1
4	7	1	6	5	3	9	8	2
2	3	6	8	7	9	1	4	5
9	8	5	1	2	4	7	6	3

296.

5	9	7	2	6	3	8	4	1
3	2	6	1	8	4	7	5	9
8	4	1	5	7	9	2	6	3
7	1	5	8	3	2	6	9	4
4	8	9	7	5	6	3	1	2
2	6	3	4	9	1	5	7	8
9	7	4	3	2	5	1	8	6
1	3	8	6	4	7	9	2	5
6	5	2	9	1	8	4	3	7

297.

2	3	7	9	1	5	8	6	4
8	9	5	3	6	4	2	7	1
1	6	4	8	2	7	5	9	3
9	8	6	2	4	1	3	5	7
4	7	2	5	3	9	1	8	6
3	5	1	7	8	6	4	2	9
6	2	3	1	7	8	9	4	5
5	4	8	6	9	3	7	1	2
7	1	9	4	5	2	6	3	8

Sudoku Hard Level

Fill the grid so that every row, every column and every 3x3 box contains the numbers 1 to 9.

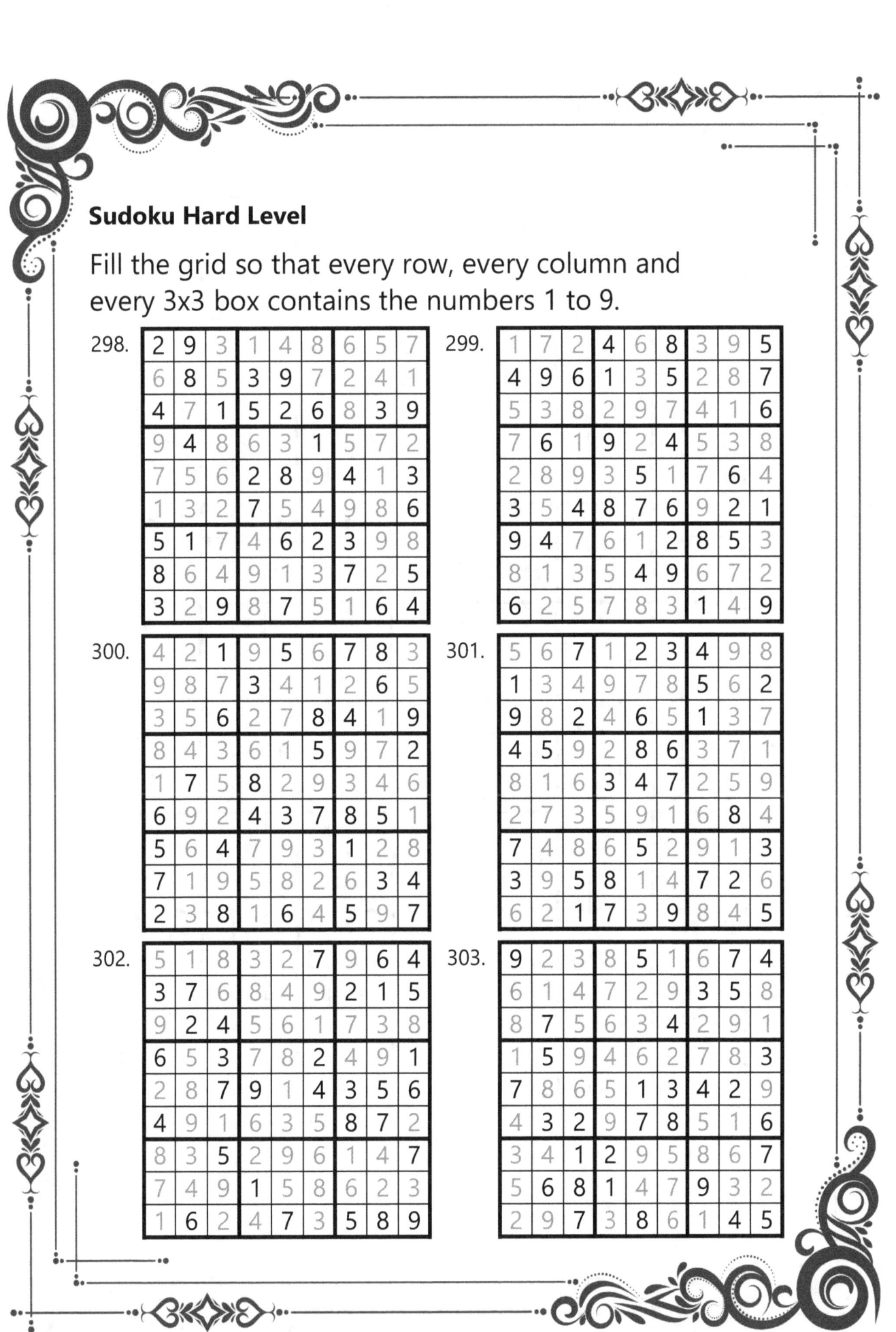

298.

299.

300.

301.

302.

303.

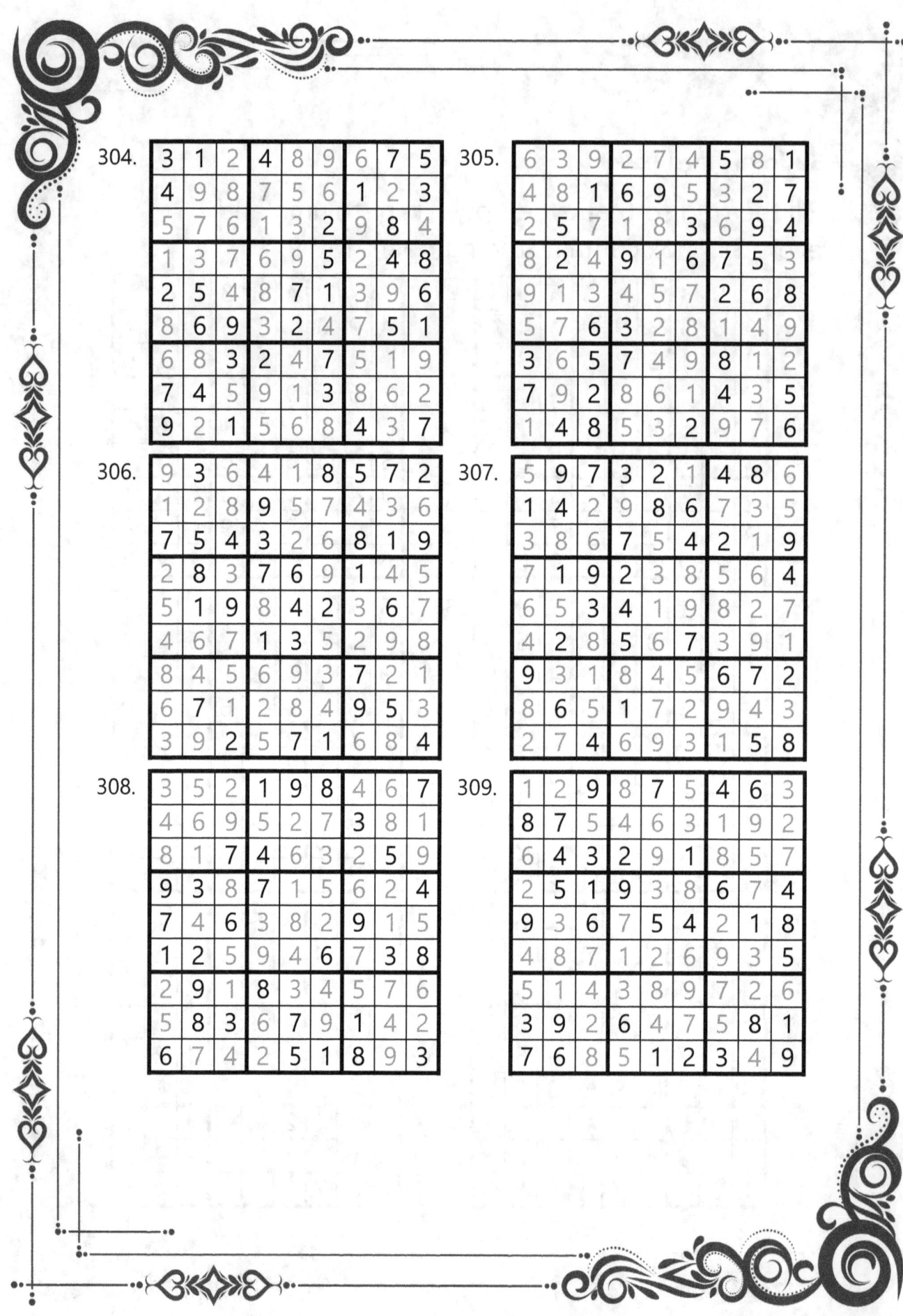

304.

3	1	2	4	8	9	6	7	5
4	9	8	7	5	6	1	2	3
5	7	6	1	3	2	9	8	4
1	3	7	6	9	5	2	4	8
2	5	4	8	7	1	3	9	6
8	6	9	3	2	4	7	5	1
6	8	3	2	4	7	5	1	9
7	4	5	9	1	3	8	6	2
9	2	1	5	6	8	4	3	7

305.

6	3	9	2	7	4	5	8	1
4	8	1	6	9	5	3	2	7
2	5	7	1	8	3	6	9	4
8	2	4	9	1	6	7	5	3
9	1	3	4	5	7	2	6	8
5	7	6	3	2	8	1	4	9
3	6	5	7	4	9	8	1	2
7	9	2	8	6	1	4	3	5
1	4	8	5	3	2	9	7	6

306.

9	3	6	4	1	8	5	7	2
1	2	8	9	5	7	4	3	6
7	5	4	3	2	6	8	1	9
2	8	3	7	6	9	1	4	5
5	1	9	8	4	2	3	6	7
4	6	7	1	3	5	2	9	8
8	4	5	6	9	3	7	2	1
6	7	1	2	8	4	9	5	3
3	9	2	5	7	1	6	8	4

307.

5	9	7	3	2	1	4	8	6
1	4	2	9	8	6	7	3	5
3	8	6	7	5	4	2	1	9
7	1	9	2	3	8	5	6	4
6	5	3	4	1	9	8	2	7
4	2	8	5	6	7	3	9	1
9	3	1	8	4	5	6	7	2
8	6	5	1	7	2	9	4	3
2	7	4	6	9	3	1	5	8

308.

3	5	2	1	9	8	4	6	7
4	6	9	5	2	7	3	8	1
8	1	7	4	6	3	2	5	9
9	3	8	7	1	5	6	2	4
7	4	6	3	8	2	9	1	5
1	2	5	9	4	6	7	3	8
2	9	1	8	3	4	5	7	6
5	8	3	6	7	9	1	4	2
6	7	4	2	5	1	8	9	3

309.

1	2	9	8	7	5	4	6	3
8	7	5	4	6	3	1	9	2
6	4	3	2	9	1	8	5	7
2	5	1	9	3	8	6	7	4
9	3	6	7	5	4	2	1	8
4	8	7	1	2	6	9	3	5
5	1	4	3	8	9	7	2	6
3	9	2	6	4	7	5	8	1
7	6	8	5	1	2	3	4	9

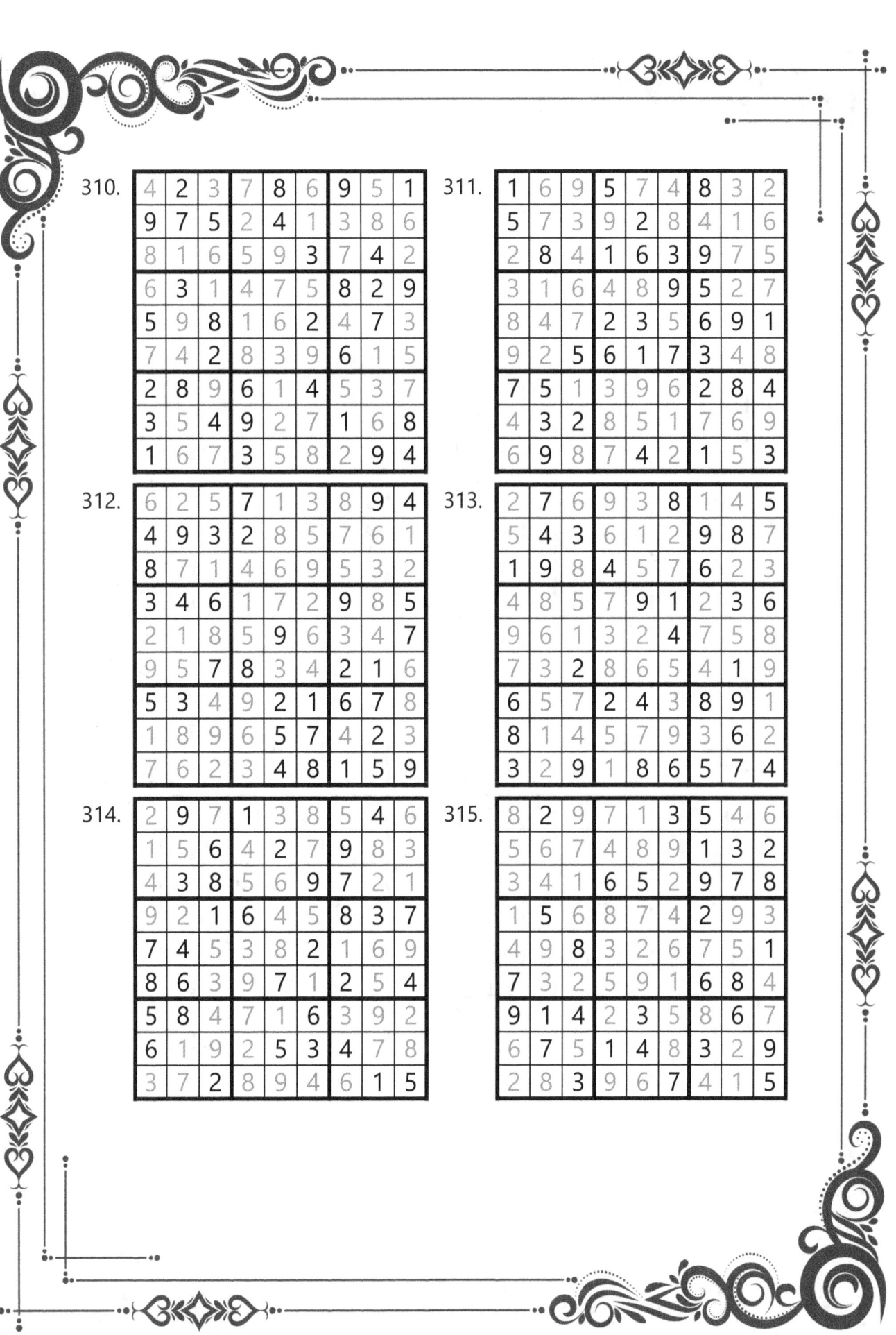

310.

4	2	3	7	8	6	9	5	1
9	7	5	2	4	1	3	8	6
8	1	6	5	9	3	7	4	2
6	3	1	4	7	5	8	2	9
5	9	8	1	6	2	4	7	3
7	4	2	8	3	9	6	1	5
2	8	9	6	1	4	5	3	7
3	5	4	9	2	7	1	6	8
1	6	7	3	5	8	2	9	4

311.

1	6	9	5	7	4	8	3	2
5	7	3	9	2	8	4	1	6
2	8	4	1	6	3	9	7	5
3	1	6	4	8	9	5	2	7
8	4	7	2	3	5	6	9	1
9	2	5	6	1	7	3	4	8
7	5	1	3	9	6	2	8	4
4	3	2	8	5	1	7	6	9
6	9	8	7	4	2	1	5	3

312.

6	2	5	7	1	3	8	9	4
4	9	3	2	8	5	7	6	1
8	7	1	4	6	9	5	3	2
3	4	6	1	7	2	9	8	5
2	1	8	5	9	6	3	4	7
9	5	7	8	3	4	2	1	6
5	3	4	9	2	1	6	7	8
1	8	9	6	5	7	4	2	3
7	6	2	3	4	8	1	5	9

313.

2	7	6	9	3	8	1	4	5
5	4	3	6	1	2	9	8	7
1	9	8	4	5	7	6	2	3
4	8	5	7	9	1	2	3	6
9	6	1	3	2	4	7	5	8
7	3	2	8	6	5	4	1	9
6	5	7	2	4	3	8	9	1
8	1	4	5	7	9	3	6	2
3	2	9	1	8	6	5	7	4

314.

2	9	7	1	3	8	5	4	6
1	5	6	4	2	7	9	8	3
4	3	8	5	6	9	7	2	1
9	2	1	6	4	5	8	3	7
7	4	5	3	8	2	1	6	9
8	6	3	9	7	1	2	5	4
5	8	4	7	1	6	3	9	2
6	1	9	2	5	3	4	7	8
3	7	2	8	9	4	6	1	5

315.

8	2	9	7	1	3	5	4	6
5	6	7	4	8	9	1	3	2
3	4	1	6	5	2	9	7	8
1	5	6	8	7	4	2	9	3
4	9	8	3	2	6	7	5	1
7	3	2	5	9	1	6	8	4
9	1	4	2	3	5	8	6	7
6	7	5	1	4	8	3	2	9
2	8	3	9	6	7	4	1	5

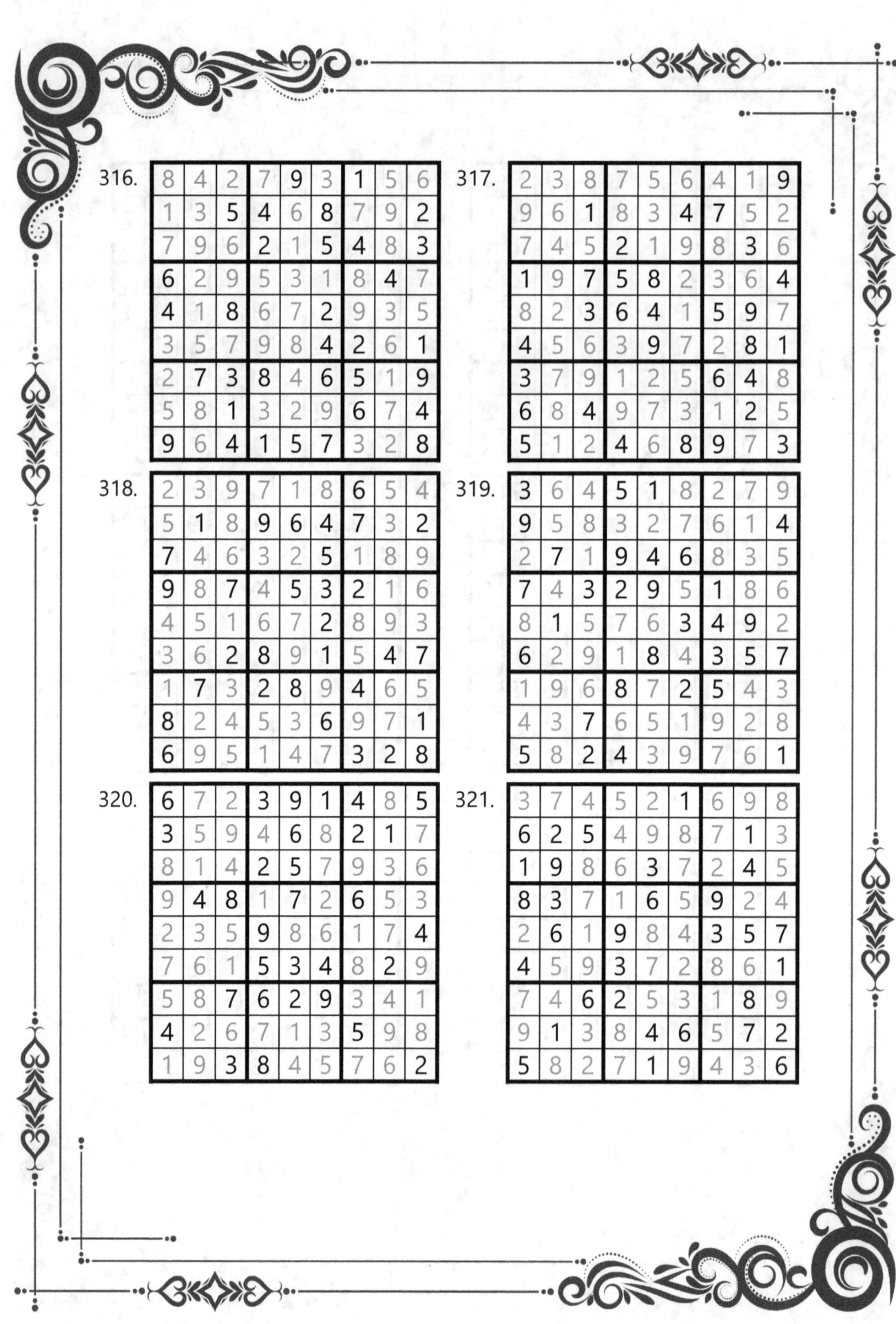

316.

8	4	2	7	9	3	1	5	6
1	3	5	4	6	8	7	9	2
7	9	6	2	1	5	4	8	3
6	2	9	5	3	1	8	4	7
4	1	8	6	7	2	9	3	5
3	5	7	9	8	4	2	6	1
2	7	3	8	4	6	5	1	9
5	8	1	3	2	9	6	7	4
9	6	4	1	5	7	3	2	8

317.

2	3	8	7	5	6	4	1	9
9	6	1	8	3	4	7	5	2
7	4	5	2	1	9	8	3	6
1	9	7	5	8	2	3	6	4
8	2	3	6	4	1	5	9	7
4	5	6	3	9	7	2	8	1
3	7	9	1	2	5	6	4	8
6	8	4	9	7	3	1	2	5
5	1	2	4	6	8	9	7	3

318.

2	3	9	7	1	8	6	5	4
5	1	8	9	6	4	7	3	2
7	4	6	3	2	5	1	8	9
9	8	7	4	5	3	2	1	6
4	5	1	6	7	2	8	9	3
3	6	2	8	9	1	5	4	7
1	7	3	2	8	9	4	6	5
8	2	4	5	3	6	9	7	1
6	9	5	1	4	7	3	2	8

319.

3	6	4	5	1	8	2	7	9
9	5	8	3	2	7	6	1	4
2	7	1	9	4	6	8	3	5
7	4	3	2	9	5	1	8	6
8	1	5	7	6	3	4	9	2
6	2	9	1	8	4	3	5	7
1	9	6	8	7	2	5	4	3
4	3	7	6	5	1	9	2	8
5	8	2	4	3	9	7	6	1

320.

6	7	2	3	9	1	4	8	5
3	5	9	4	6	8	2	1	7
8	1	4	2	5	7	9	3	6
9	4	8	1	7	2	6	5	3
2	3	5	9	8	6	1	7	4
7	6	1	5	3	4	8	2	9
5	8	7	6	2	9	3	4	1
4	2	6	7	1	3	5	9	8
1	9	3	8	4	5	7	6	2

321.

3	7	4	5	2	1	6	9	8
6	2	5	4	9	8	7	1	3
1	9	8	6	3	7	2	4	5
8	3	7	1	6	5	9	2	4
2	6	1	9	8	4	3	5	7
4	5	9	3	7	2	8	6	1
7	4	6	2	5	3	1	8	9
9	1	3	8	4	6	5	7	2
5	8	2	7	1	9	4	3	6

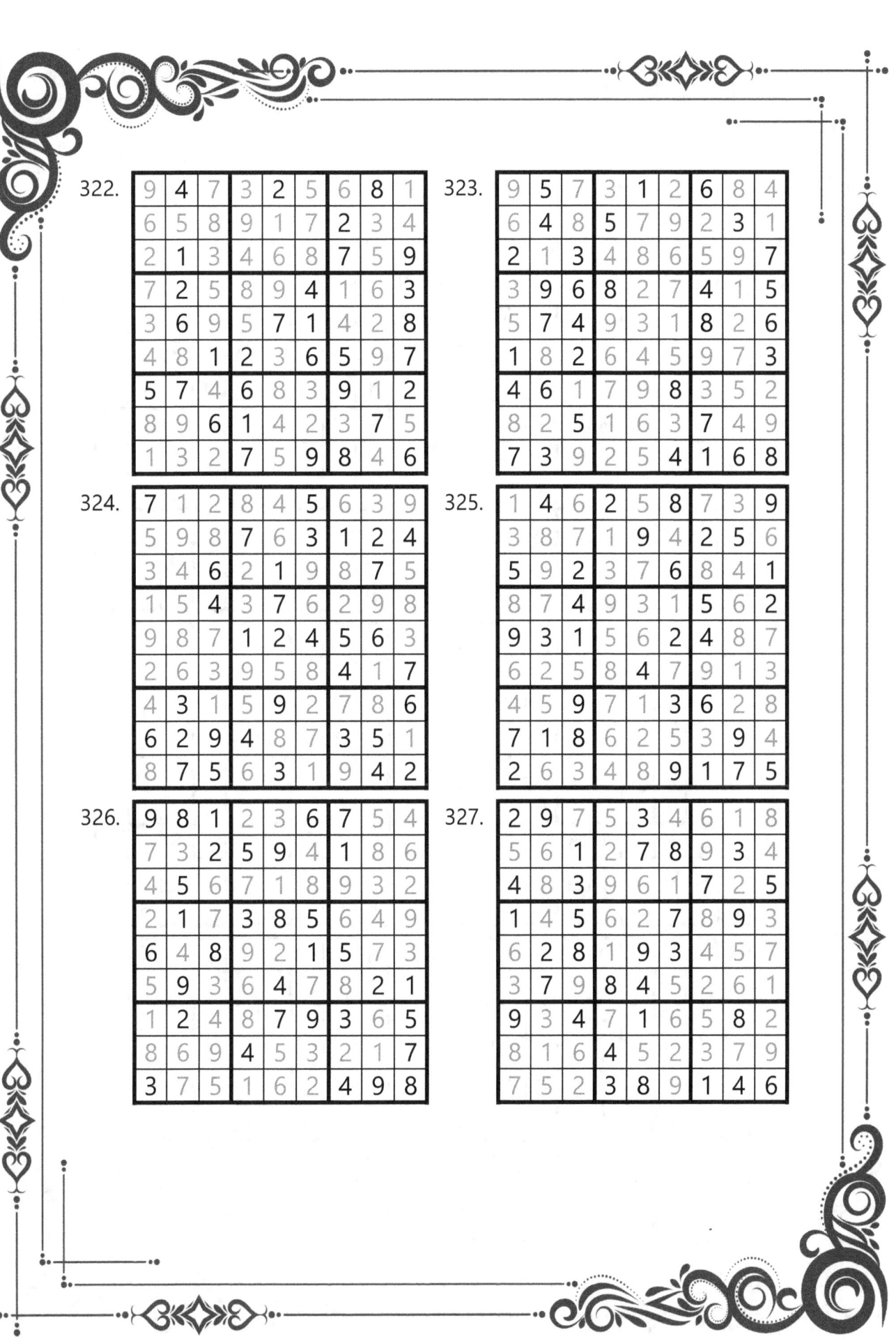

322.

9	4	7	3	2	5	6	8	1
6	5	8	9	1	7	2	3	4
2	1	3	4	6	8	7	5	9
7	2	5	8	9	4	1	6	3
3	6	9	5	7	1	4	2	8
4	8	1	2	3	6	5	9	7
5	7	4	6	8	3	9	1	2
8	9	6	1	4	2	3	7	5
1	3	2	7	5	9	8	4	6

323.

9	5	7	3	1	2	6	8	4
6	4	8	5	7	9	2	3	1
2	1	3	4	8	6	5	9	7
3	9	6	8	2	7	4	1	5
5	7	4	9	3	1	8	2	6
1	8	2	6	4	5	9	7	3
4	6	1	7	9	8	3	5	2
8	2	5	1	6	3	7	4	9
7	3	9	2	5	4	1	6	8

324.

7	1	2	8	4	5	6	3	9
5	9	8	7	6	3	1	2	4
3	4	6	2	1	9	8	7	5
1	5	4	3	7	6	2	9	8
9	8	7	1	2	4	5	6	3
2	6	3	9	5	8	4	1	7
4	3	1	5	9	2	7	8	6
6	2	9	4	8	7	3	5	1
8	7	5	6	3	1	9	4	2

325.

1	4	6	2	5	8	7	3	9
3	8	7	1	9	4	2	5	6
5	9	2	3	7	6	8	4	1
8	7	4	9	3	1	5	6	2
9	3	1	5	6	2	4	8	7
6	2	5	8	4	7	9	1	3
4	5	9	7	1	3	6	2	8
7	1	8	6	2	5	3	9	4
2	6	3	4	8	9	1	7	5

326.

9	8	1	2	3	6	7	5	4
7	3	2	5	9	4	1	8	6
4	5	6	7	1	8	9	3	2
2	1	7	3	8	5	6	4	9
6	4	8	9	2	1	5	7	3
5	9	3	6	4	7	8	2	1
1	2	4	8	7	9	3	6	5
8	6	9	4	5	3	2	1	7
3	7	5	1	6	2	4	9	8

327.

2	9	7	5	3	4	6	1	8
5	6	1	2	7	8	9	3	4
4	8	3	9	6	1	7	2	5
1	4	5	6	2	7	8	9	3
6	2	8	1	9	3	4	5	7
3	7	9	8	4	5	2	6	1
9	3	4	7	1	6	5	8	2
8	1	6	4	5	2	3	7	9
7	5	2	3	8	9	1	4	6

Sudoku Expert Level

Fill the grid so that every row, every column and every 3x3 box contains the numbers 1 to 9.

328.   329.

330.   331.

332.   333.

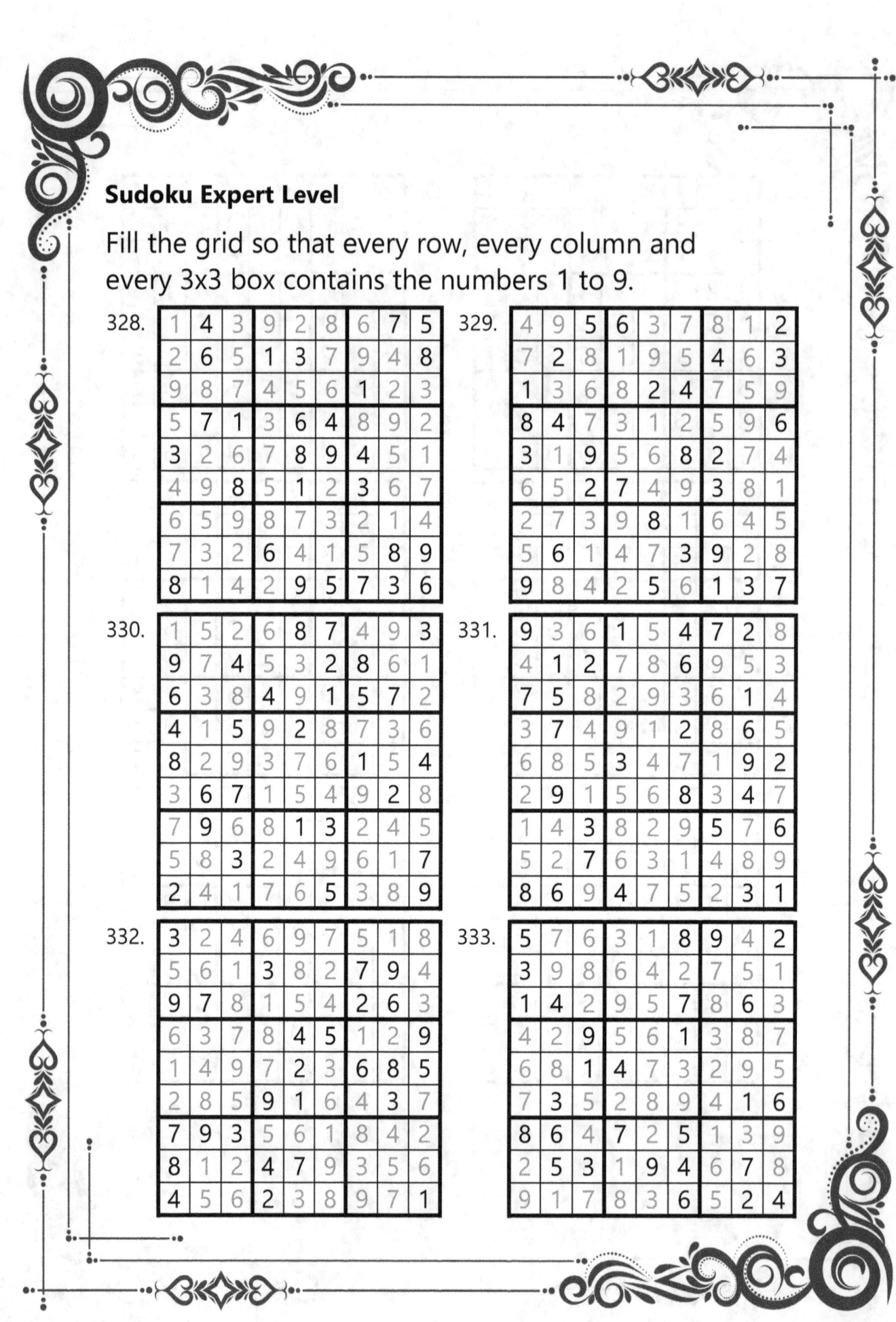

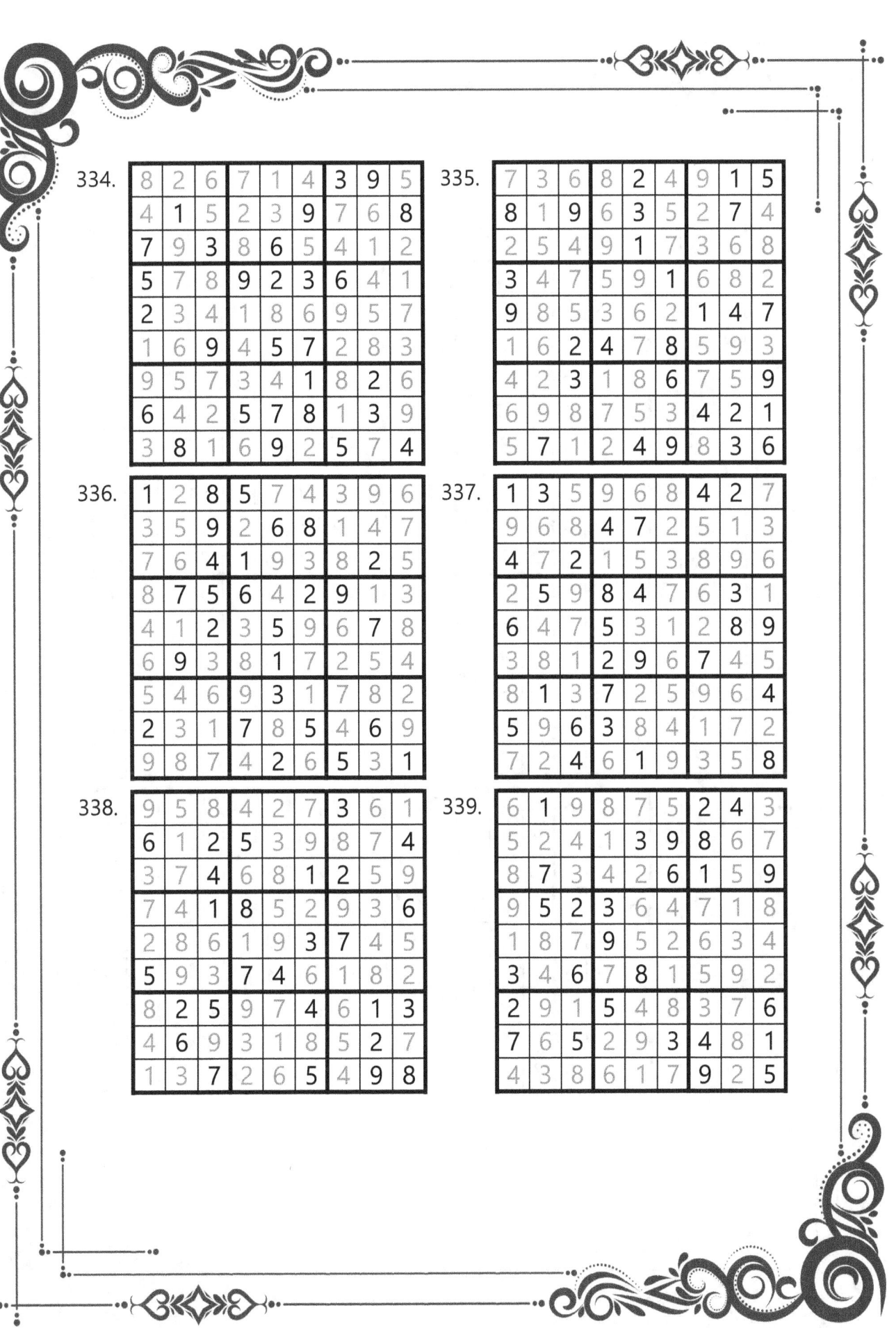

334.

8	2	6	7	1	4	3	9	5
4	1	5	2	3	9	7	6	8
7	9	3	8	6	5	4	1	2
5	7	8	9	2	3	6	4	1
2	3	4	1	8	6	9	5	7
1	6	9	4	5	7	2	8	3
9	5	7	3	4	1	8	2	6
6	4	2	5	7	8	1	3	9
3	8	1	6	9	2	5	7	4

335.

7	3	6	8	2	4	9	1	5
8	1	9	6	3	5	2	7	4
2	5	4	9	1	7	3	6	8
3	4	7	5	9	1	6	8	2
9	8	5	3	6	2	1	4	7
1	6	2	4	7	8	5	9	3
4	2	3	1	8	6	7	5	9
6	9	8	7	5	3	4	2	1
5	7	1	2	4	9	8	3	6

336.

1	2	8	5	7	4	3	9	6
3	5	9	2	6	8	1	4	7
7	6	4	1	9	3	8	2	5
8	7	5	6	4	2	9	1	3
4	1	2	3	5	9	6	7	8
6	9	3	8	1	7	2	5	4
5	4	6	9	3	1	7	8	2
2	3	1	7	8	5	4	6	9
9	8	7	4	2	6	5	3	1

337.

1	3	5	9	6	8	4	2	7
9	6	8	4	7	2	5	1	3
4	7	2	1	5	3	8	9	6
2	5	9	8	4	7	6	3	1
6	4	7	5	3	1	2	8	9
3	8	1	2	9	6	7	4	5
8	1	3	7	2	5	9	6	4
5	9	6	3	8	4	1	7	2
7	2	4	6	1	9	3	5	8

338.

9	5	8	4	2	7	3	6	1
6	1	2	5	3	9	8	7	4
3	7	4	6	8	1	2	5	9
7	4	1	8	5	2	9	3	6
2	8	6	1	9	3	7	4	5
5	9	3	7	4	6	1	8	2
8	2	5	9	7	4	6	1	3
4	6	9	3	1	8	5	2	7
1	3	7	2	6	5	4	9	8

339.

6	1	9	8	7	5	2	4	3
5	2	4	1	3	9	8	6	7
8	7	3	4	2	6	1	5	9
9	5	2	3	6	4	7	1	8
1	8	7	9	5	2	6	3	4
3	4	6	7	8	1	5	9	2
2	9	1	5	4	8	3	7	6
7	6	5	2	9	3	4	8	1
4	3	8	6	1	7	9	2	5

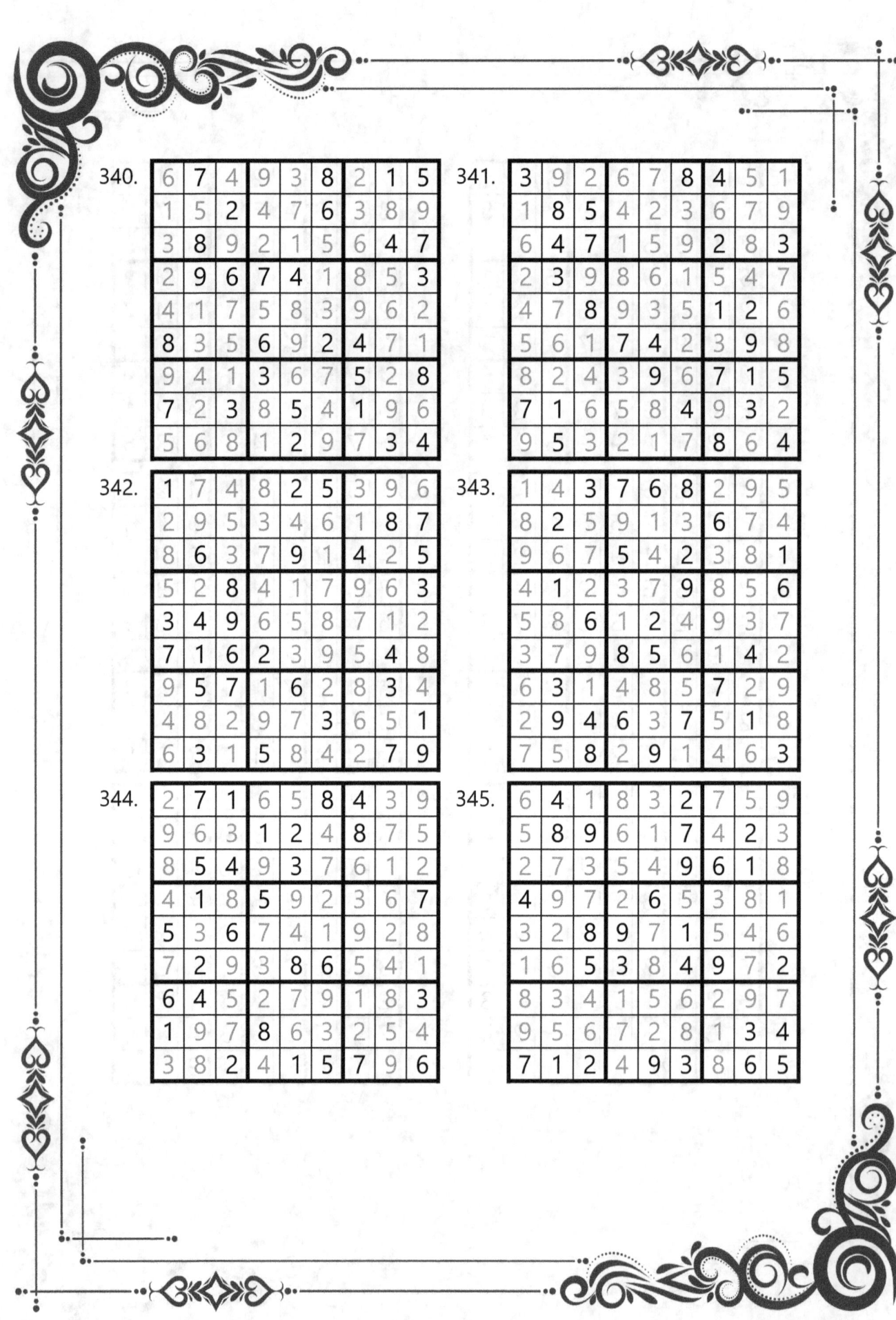

340.

6	7	4	9	3	8	2	1	5
1	5	2	4	7	6	3	8	9
3	8	9	2	1	5	6	4	7
2	9	6	7	4	1	8	5	3
4	1	7	5	8	3	9	6	2
8	3	5	6	9	2	4	7	1
9	4	1	3	6	7	5	2	8
7	2	3	8	5	4	1	9	6
5	6	8	1	2	9	7	3	4

341.

3	9	2	6	7	8	4	5	1
1	8	5	4	2	3	6	7	9
6	4	7	1	5	9	2	8	3
2	3	9	8	6	1	5	4	7
4	7	8	9	3	5	1	2	6
5	6	1	7	4	2	3	9	8
8	2	4	3	9	6	7	1	5
7	1	6	5	8	4	9	3	2
9	5	3	2	1	7	8	6	4

342.

1	7	4	8	2	5	3	9	6
2	9	5	3	4	6	1	8	7
8	6	3	7	9	1	4	2	5
5	2	8	4	1	7	9	6	3
3	4	9	6	5	8	7	1	2
7	1	6	2	3	9	5	4	8
9	5	7	1	6	2	8	3	4
4	8	2	9	7	3	6	5	1
6	3	1	5	8	4	2	7	9

343.

1	4	3	7	6	8	2	9	5
8	2	5	9	1	3	6	7	4
9	6	7	5	4	2	3	8	1
4	1	2	3	7	9	8	5	6
5	8	6	1	2	4	9	3	7
3	7	9	8	5	6	1	4	2
6	3	1	4	8	5	7	2	9
2	9	4	6	3	7	5	1	8
7	5	8	2	9	1	4	6	3

344.

2	7	1	6	5	8	4	3	9
9	6	3	1	2	4	8	7	5
8	5	4	9	3	7	6	1	2
4	1	8	5	9	2	3	6	7
5	3	6	7	4	1	9	2	8
7	2	9	3	8	6	5	4	1
6	4	5	2	7	9	1	8	3
1	9	7	8	6	3	2	5	4
3	8	2	4	1	5	7	9	6

345.

6	4	1	8	3	2	7	5	9
5	8	9	6	1	7	4	2	3
2	7	3	5	4	9	6	1	8
4	9	7	2	6	5	3	8	1
3	2	8	9	7	1	5	4	6
1	6	5	3	8	4	9	7	2
8	3	4	1	5	6	2	9	7
9	5	6	7	2	8	1	3	4
7	1	2	4	9	3	8	6	5

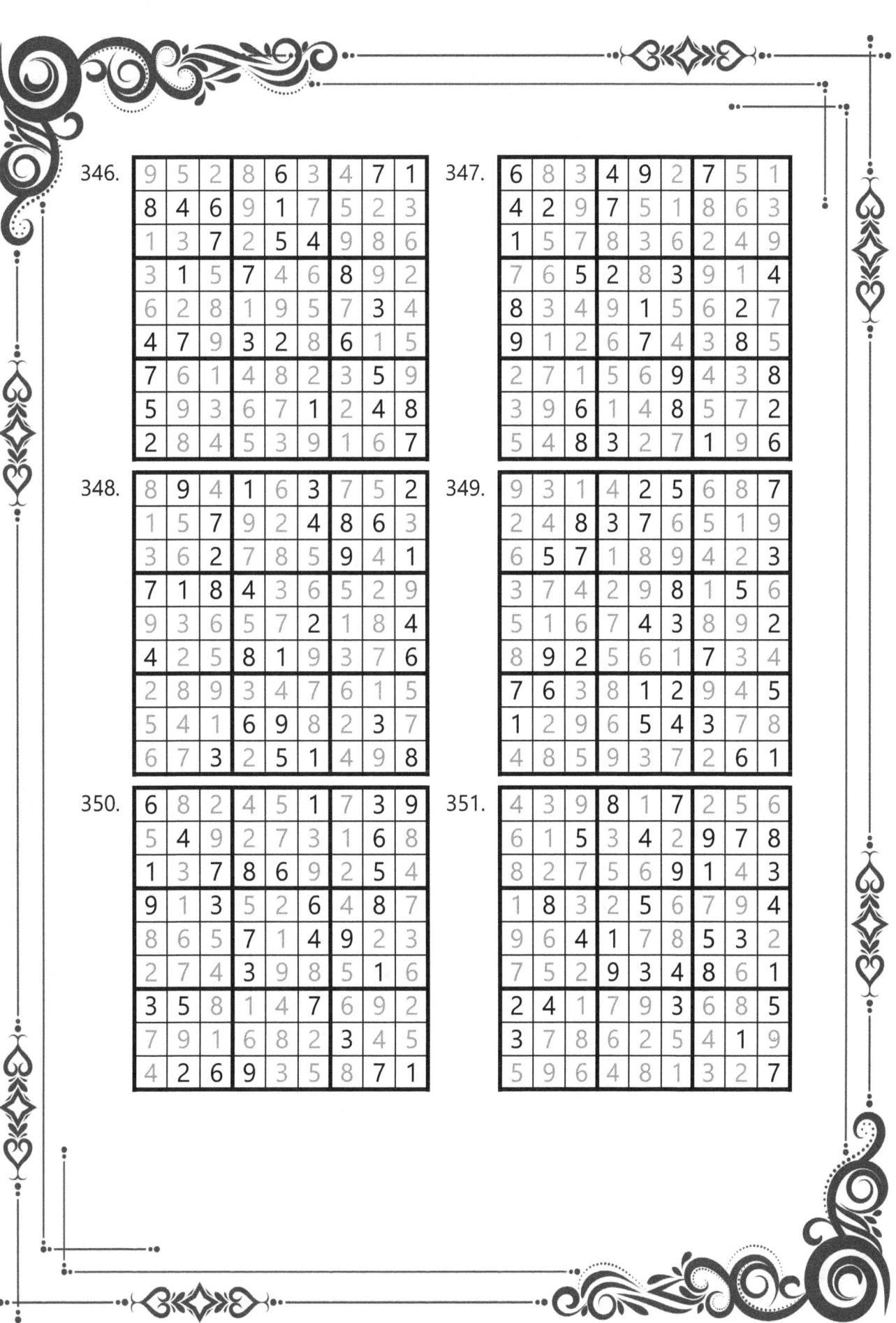

346.

9	5	2	8	6	3	4	7	1
8	4	6	9	1	7	5	2	3
1	3	7	2	5	4	9	8	6
3	1	5	7	4	6	8	9	2
6	2	8	1	9	5	7	3	4
4	7	9	3	2	8	6	1	5
7	6	1	4	8	2	3	5	9
5	9	3	6	7	1	2	4	8
2	8	4	5	3	9	1	6	7

347.

6	8	3	4	9	2	7	5	1
4	2	9	7	5	1	8	6	3
1	5	7	8	3	6	2	4	9
7	6	5	2	8	3	9	1	4
8	3	4	9	1	5	6	2	7
9	1	2	6	7	4	3	8	5
2	7	1	5	6	9	4	3	8
3	9	6	1	4	8	5	7	2
5	4	8	3	2	7	1	9	6

348.

8	9	4	1	6	3	7	5	2
1	5	7	9	2	4	8	6	3
3	6	2	7	8	5	9	4	1
7	1	8	4	3	6	5	2	9
9	3	6	5	7	2	1	8	4
4	2	5	8	1	9	3	7	6
2	8	9	3	4	7	6	1	5
5	4	1	6	9	8	2	3	7
6	7	3	2	5	1	4	9	8

349.

9	3	1	4	2	5	6	8	7
2	4	8	3	7	6	5	1	9
6	5	7	1	8	9	4	2	3
3	7	4	2	9	8	1	5	6
5	1	6	7	4	3	8	9	2
8	9	2	5	6	1	7	3	4
7	6	3	8	1	2	9	4	5
1	2	9	6	5	4	3	7	8
4	8	5	9	3	7	2	6	1

350.

6	8	2	4	5	1	7	3	9
5	4	9	2	7	3	1	6	8
1	3	7	8	6	9	2	5	4
9	1	3	5	2	6	4	8	7
8	6	5	7	1	4	9	2	3
2	7	4	3	9	8	5	1	6
3	5	8	1	4	7	6	9	2
7	9	1	6	8	2	3	4	5
4	2	6	9	3	5	8	7	1

351.

4	3	9	8	1	7	2	5	6
6	1	5	3	4	2	9	7	8
8	2	7	5	6	9	1	4	3
1	8	3	2	5	6	7	9	4
9	6	4	1	7	8	5	3	2
7	5	2	9	3	4	8	6	1
2	4	1	7	9	3	6	8	5
3	7	8	6	2	5	4	1	9
5	9	6	4	8	1	3	2	7